T0091933

WATER

WATER

A Visual and Scientific History

Jack Challoner

The MIT Press
Cambridge, Massachusetts
London, England

The MIT Press would like to thank the anonymous peer reviewers who provided comments on drafts of this book. The generous work of academic experts is essential for establishing the authority and quality of our publications. We acknowledge with gratitude the contributions of these otherwise uncredited readers.

This book was set in Gazette and Foco by the MIT Press. Printed and bound in the United States of America.

Library of Congress Cataloging-in-Publication Data
Names: Challoner, Jack, author.
Title: Water : a visual and scientific history / Jack Challoner.
Description: Cambridge, Massachusetts : The MIT Press, [2020] | Includes bibliographical references and index.
Identifiers: LCCN 2020047201 | ISBN 9780262046145 (hardcover)
Subjects: LCSH: Water. | Water—Analysis. | Water chemistry.
Classification: LCC QD169.W3 C42 2020 | DDC 553.7—dc23
LC record available at https://lccn.loc.gov/2020047201

10 9 8 7 6 5 4 3 2 1

Contents

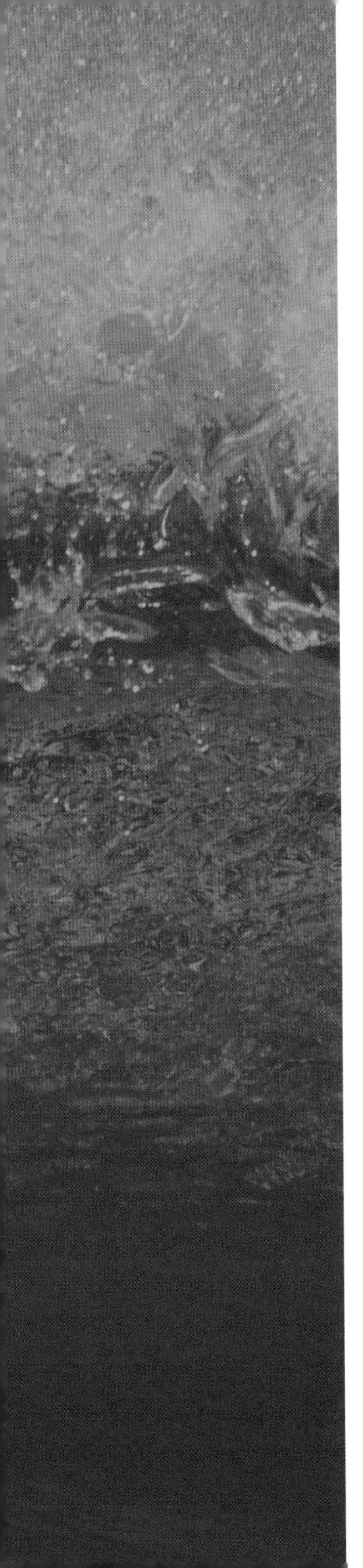

Introduction

In her 1995 book *Sea Change: A Message of the Oceans*, American oceanographer Sylvia Earle wrote: "There's plenty of water in the Universe without life, but nowhere is there life without water." She is right on both counts. Chapter 1 of this book examines the origins of water in the Universe, and where it can be found out there in space. Chapter 2 considers the question of how water came to be on Earth, and follows water on its relentless journey around our planet's water cycle—including the vital role water played in the origins of human civilization.

Water is so familiar and important it is easy to take it for granted. The average American uses 90 gallons of water every day—mostly for washing and cooking, and to flush away waste. Nearly every liquid we encounter in our everyday lives is water, on its own or with other substances dissolved or suspended in it. Milk, for example, is 87 percent water, with fats, proteins and sugars making up the rest. Water is "embedded" in every product we buy—nearly 2,000 gallons go into producing every pound of beef, and between 10,000 and 20,000 gallons to make a car. Water plays a central role in our climate: clouds distribute heat around the world, oceans are a major sink of carbon dioxide, and huge volumes of ice at the poles mitigate against fluctuations in temperature.

Chapter 3 traces the history of how great thinkers came to discover what water is made of, and reveals the interactions of water molecules with each other. These interactions underlie the existence and behaviors of water in its solid, liquid and vapor form—which are explored in detail in chapter 4.

Water is so important and so familiar that it would be reasonable to assume that it is an "ordinary" liquid. But in fact, it is a liquid like no other liquid—it has a unique set of properties. So simple and small a molecule, and so well studied, is water that it would be reasonable to assume we understand it completely. And yet much remains unfathomed about this unique substance. The properties of water have their greatest influence at interfaces, between water and other substances—as explored in chapter 5. And, as Sylvia Earle wrote, "nowhere is there life without water." Chapter 6 explores how water's interactions with other substances make life possible—here on Earth, and perhaps elsewhere.

The smallest droplets in ocean spray are about half of one-thousandth of an inch in diameter. So small are the water molecules that a droplet of that size is made of more than a trillion of them. *Source:* Roger Mosley.

CHAPTER 1

"Water, Water, Every Where"

The title of this chapter (including the unusual spelling of *every-where*) comes from Samuel Taylor Coleridge's poem "The Rime of the Ancient Mariner." In the poem, the eponymous mariner tells of how winds carried his ship from the tumultuous, icy waters of the Southern Ocean to calmer, warm waters nearer the equator. There, the ship became stuck, barely drifting in almost nonexistent winds. The mariner reflects that while the ship was stranded, there was water all around "nor any drop to drink"—and Coleridge stresses our fundamental need for water in many of the poem's verses. In a sense, we too find ourselves drifting in a vast ocean of water, for H_2O is one of the most abundant compounds in the Universe. And, like the ocean on which the mariner's ship was drifting, most of this cosmic ocean is not the familiar clean liquid on which our lives depend. So what kind of water is out there in space? And when and how did it form?

In the Beginning

It is perhaps no surprise that water is so common in the cosmos: it is made of hydrogen—by far the most abundant element—and oxygen, the third most abundant. Hydrogen was created in the first minutes of the Universe, around 13.8 billion years ago. It was hydrogen nuclei, not atoms of hydrogen, that came into being in that earliest time. This is because the Universe was so hot that electrons were too energetic to attach to nuclei—and that attachment is a necessary prerequisite for atoms. Most hydrogen nuclei are simply naked protons; a small proportion have a neutron attached, making a nucleus of hydrogen-2, deuterium—which, as we shall see, is very important in the story of water in our Solar System. Also during those first minutes, some of the hydrogen and deuterium nuclei that had been created joined together to make nuclei of helium (and a much smaller number of lithium nuclei). Around 300,000 years after the Big Bang, the Universe had cooled enough for electrons to attach to nuclei: the first atoms, of hydrogen (including some deuterium), helium and lithium, finally came to be.

To this day, hydrogen and helium make up more than 99 percent of the matter in the Universe (not counting dark matter, which outweighs normal matter by a factor of more than eight—but that is another, as yet poorly understood, story). The hydrogen, deuterium and helium nuclei were the starting points for the creation of the other elements, including oxygen, in energetic nuclear reactions—mostly inside stars and in their violent death throes, supernovas. The first stars flickered on around 200 million years after the Big Bang, as huge clouds of hydrogen and helium gas collapsed in clumps, creating pressures and temperatures high enough for those element-building nuclear reactions to begin at their cores. The first oxygen nuclei would have been created soon after the first stars were born.

The Horsehead Nebula, about 1,600 light years away. The characteristic red color is from hydrogen gas, energized by ultraviolet radiation from young stars. The illuminated cloud is partly obscured by a dark, non-star-forming cloud. *Source:* National Optical Astronomy Observatory/Association of Universities for Research in Astronomy/National Science Foundation.

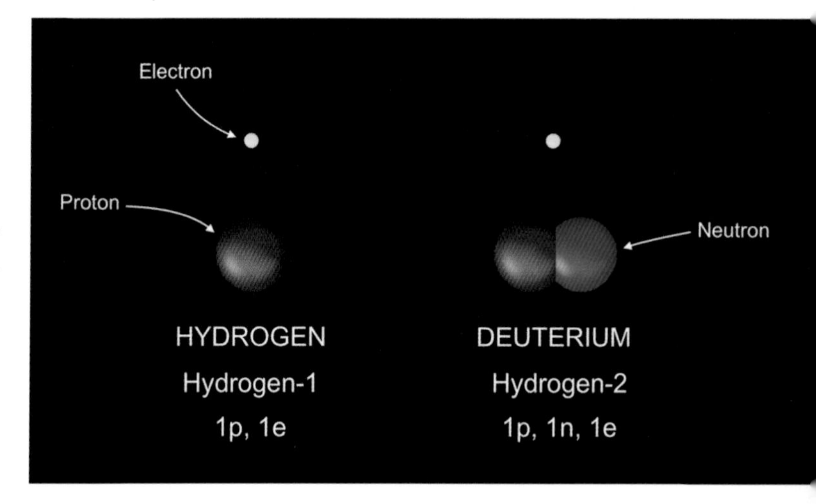

Nuclei of hydrogen and deuterium, plus an electron for each. The proton (purple) carries a positive charge and the electron (yellow) carries the equivalent amount of negative charge, while the neutron (blue) is neutral—so an atom is neutral overall. For the first 300,000 years, the Universe was too hot for electrons to attach to the nuclei.

No water molecules could form until the oxygen created in the cores of those first stars was released. Supernovas spilled the dying stars' contents across immense volumes of space in unimaginably powerful explosions, forming vast clouds of gas. In 2017, a team of astronomers using the Atacama Large Millimeter/submillimeter Array in Chile detected the characteristic "fingerprint" of oxygen molecules in the radiation coming from such interstellar gas clouds inside an extremely distant galaxy.[1] So far away is the galaxy that the radiation the team detected had traveled across space for more than 13 billion years. The astronomers estimated that the oxygen would have been released into space around 600 million years after the Big Bang. Now that both hydrogen and oxygen atoms existed; water had at last become a possibility.

Making Water

Astrophysicists have identified several different reaction pathways by which water molecules form in space. In cool interstellar gas clouds scattered across space by supernovas, the most prevalent reaction involves ionized (electrically charged) hydrogen molecules and oxygen atoms. It begins with the formation of H_3^+ ions, as neutral hydrogen molecules (H_2) are pummeled by energetic cosmic rays (the extra H is a positively charged cosmic ray proton, a hydrogen nucleus). The H_3^+ ions react with oxygen atoms in the gas, forming OH^+ ions and molecular hydrogen. After a couple more molecular-ionic exchanges, the result is familiar, neutral (uncharged) molecules of water. The H_3^+ ions that take part in these reactions are more likely to contain atoms of deuterium than the hydrogen gas at large (the reason for this is connected to the fact that the deuterium atom has more mass than an ordinary hydrogen atom). And since the collection of H_3^+ ions becomes enriched with deuterium atoms, so too does the water formed in cold interstellar gas clouds. The ratio of deuterium atoms to hydrogen atoms (D/H) in water molecules in interstellar gas clouds is typically around 10,000 parts per million. By comparison, the D/H ratio in primordial hydrogen gas is only about 16 ppm.

Artist's impression of torus of material feeding a huge black hole at the center of the galaxy APM 08279+5255, where scientists found the earliest known and largest reservoir of water in the Universe. *Source:* NASA/ESA.

The main reaction by which water is made in gas clouds involves molecules and ions.

H^+ (proton)

$$H_2 + H^+ \longrightarrow H_3^+$$

$$H_3^+ + O \longrightarrow OH^+ + H_2$$

$$OH^+ + H_2 \longrightarrow H_2O^+ + H$$

$$H_2O^+ + H_2 \longrightarrow H_3O^+ + H$$

e^- (electron)

$$H_3O^+ + e^- \longrightarrow H_2O + H$$

Interstellar gas clouds contain tiny grains of *dust*, each with a diameter less than one-hundredth the width of a human hair. Even in the densest regions of these clouds, in which stars form, the material is extremely tenuous: a volume of space the size of a football stadium would contain only one or two dust grains, and each cubic inch only about 10,000 atoms and molecules. (By contrast, the air you are breathing has a billion trillion atoms and molecules per cubic inch.) Dust grains provide a surface on which water vapor can condense—and on which more water molecules can form. Atoms and molecules of hydrogen and oxygen land on a dust grain's surface and self-assemble into H_2O molecules, like tiny autonomous Legos. This reaction also favors deuterium and contributes to the raised D/H ratio in interstellar clouds. In 2010, French and Italian physicists created conditions like those on an interstellar dust grain inside their laboratory—and hydrogen and oxygen self-assembled into water molecules, as theory had predicted.[2] The outer regions of dense stellar nebulas offer protection for water molecules deeper inside. American astronomer Charles Townes wrote in 2006: "That much gas, plus dust located in the same interstellar cloud, was adequate to shield molecules in the inner parts of the cloud from ultraviolet radiation that would otherwise tear them apart."[3]

Charles Townes provided the first evidence of the existence of water in interstellar space. In the 1950s, he had suggested that small molecules such as ammonia and water ought to exist in gas clouds and that radio telescopes might be the best way to find them. He and his team found both of these compounds in 1969, by picking up characteristic frequencies of the microwave radiation they had emitted. Since then, astronomers have detected water in space throughout our galaxy—and in many other galaxies besides. In 2011, two teams of astronomers discovered a water-rich region around a black hole at the center of a distant galaxy, called APM 08279+5255—in this case by studying radiation that was produced about 1.6 billion years after the Big Bang. The amount of water vapor there, once condensed, would fill Earth's oceans more than 100 trillion times over.[4] This is the biggest reservoir of water so far detected, and one of the oldest (see page 5).

In the cool, diffuse clouds of gas in the interstellar space of our own galaxy, one in every thousand or so oxygen atoms, and one in a hundred million or so hydrogen atoms, are bound up in water molecules.[5] In slightly warmer gas clouds, particularly where powerful shock waves push through and increase the cloud's density, neutral oxygen atoms and neutral hydrogen molecules react directly[6]—though not quite in the same way as when hydrogen burns in oxygen-rich air here on Earth. In these places, water is much more abundant, as nearly all the available oxygen becomes bound to hydrogen atoms.

Water in the System

Gravity causes dense regions of interstellar gas to collapse in on themselves, creating star-forming regions called nebulas. Around 4.6 billion years ago, the star we call the Sun formed in a nebula, in one arm of a galaxy we call the Milky Way. In the *protoplanetary disk* surrounding the young Sun, the solid material, including plenty of water ice, clumped together, sticking after colliding—this process is called *accretion*. As a particular object grows more massive, its rate of accretion increases, because it is a larger "target" for smaller pieces of material, and because of its increasing gravitational attraction as its mass increases. This is how the planets and their moons, minor planets (rocky objects including asteroids) and comets of the Solar System came to be. Water has been detected and studied on all of these objects. Surprisingly, perhaps, there is even water on the Sun. While the majority of the Sun is far too hot for water to exist, surface features known as sunspots are significantly cooler—albeit still a roasting 6,500 degrees Fahrenheit (°F). Astronomers first observed hot water in the spectrum of sunspots in 1995.[7]

Studying the ratio of deuterium to ordinary hydrogen in the water on planets, moons, minor planets, and comets provides crucial evidence in the quest to unravel the story of our system's water in general, and the source of Earth's water in particular. This is useful because the D/H ratio of the Solar System's water would have varied, across space and time, as the Solar System formed and matured. The D/H ratio of water in the cold gas cloud from which the Sun was born was very high, thanks to those reactions that formed the water, which favor deuterium, as described above. But near the young Sun, the ratio changed dramatically. At higher temperatures, ordinary hydrogen atoms tended to replace deuterium in water molecules—a reaction called *isotopic exchange*. As a result, the hydrogen in water molecules close to the Sun became *equilibriated* with hydrogen gas. The D/H ratio of water there was reduced from around 10,000 parts per million to a figure close to that for hydrogen in the Universe at large—around 20 parts per million. Isotopic exchange happened less frequently at lower temperatures, so the D/H ratio for water further out remained higher.

For the first few hundred million miles out from the Sun, up to an imaginary line called the *snow line* (or frost line), the intensity of the solar radiation means that water could exist only as vapor (and beyond which, it could freeze out as ice). The newly formed lower-D/H water vapor mixed with the undisturbed higher-D/H

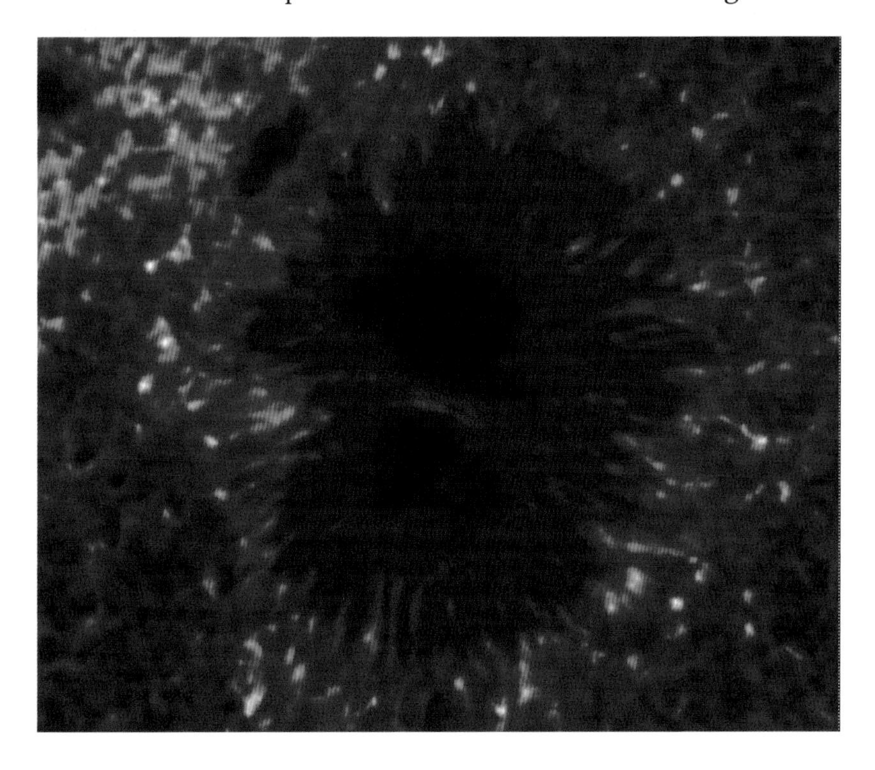

This remarkable image of a sunspot is a combination of optical, X-ray, ultraviolet and infrared data. Very hot water is abundant in sunspots, which are typically many times the size of Earth. *Source:* NASA.

water vapor; this mixing, along with the reduction in temperature with distance from the Sun, led to the creation of a D/H gradient. Water's D/H ratio was lowest close to the Sun and highest further out toward the snow line in the undisturbed water. Water in different parts of the Solar System today still tends to reflect this gradient, which formed within a million years or so after the creation of the Sun. However, processes such as the loss of planetary atmospheres, objects colliding and merging or changing their orbits, and the melting of water-bearing rocks can alter the D/H ratio of a particular planet or moon. These processes make it more challenging to unpick the story of the Solar System's water.

The rest of this chapter will explore some of the places where water has been found in our Solar System, starting with objects that have long been known to hold water: comets.

Comets: Dirty Snowballs

The objects in the Solar System whose orbits take them furthest from the Sun are the long period comets: those comets whose orbits take longer than 200 years to complete. A comet, often described as a "dirty snowball," is a dusty, rocky object with plenty of water, locked away as ice. Despite the image the epithet "dirty snowball" conjures up, comets are not white with chunks of dark material embedded in them: they are very dark in color, and most of the ice is beneath the surface, more like a snowy dirtball. The dark surface absorbs most of the sunlight that falls on it, so when and if a comet's orbit takes it close to the Sun, within the snow line, the comet heats up. The water ice *sublimes* (turns directly to a vapor), along with its other volatile materials, giving the comet a thin atmosphere, called a *coma*. The water and other light materials released are pushed into a long, straight tail by the *solar wind*—a stream of energetic particles emanating from the Sun's surface—so that the tail points away from the Sun. The ultraviolet radiation and the solar wind particles will break apart (dissociate) most of the water in the tail. Heavier dust particles

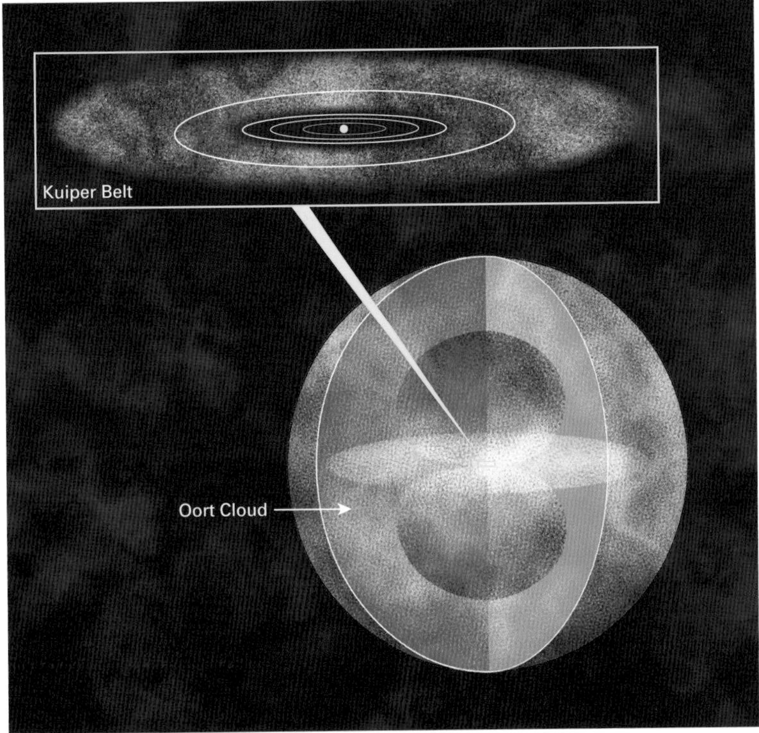

Water and dust spill from comet 67P/Churyumov-Gerasimenko, photographed by the European Space Agency's Rosetta spacecraft, from just 20 miles above the comet's surface. *Source:* European Space Agency.

Illustration showing the sizes and locations of the spherical Oort Cloud and the disk-shaped Kuiper Belt, inset. Just inside the Kuiper Belt are the orbits of the Solar System's planets. *Source:* Based on an image by NASA.

The first ever confirmed active comet that originated in another star system, comet 2I/Borisov began releasing water at a rate of about 8.5 gallons per second as it neared our Sun, in December 2019. *Source:* NASA/JPL.

are also released, and they form a separate tail, a trail of debris that is not pushed into a straight line by the solar wind. The particles in the debris are the source of most of the meteors (shooting stars) that burn up high in Earth's atmosphere when our planet crosses the trail of debris left in a comet's wake.

Most long period comets originate in the Oort Cloud, a huge spherical shell of icy objects at the edge of the Solar System. Distances in the Solar System are often measured in *astronomical units* (AU); 1 AU is the average distance between the Sun and Earth, about 93 million miles. The inner edge of the Oort Cloud lies a dizzying 2,000 AU from the Sun, the outer edge tens of thousands of AU. The Oort Cloud is likely home to around a trillion icy chunks of material. Shorter period comets come from the Kuiper Belt, which begins about 30 AU and ends about 100 AU from the Sun. In 2019, astronomers for the first time spotted a comet inside our Solar System that originated in a different star system. As it neared the Sun, comet 2I/Borisov began losing water—as one of the scientists studying it wrote, this comet was releasing alien water into our neighborhood.[8]

The water of long period comets has the highest D/H ratio of any objects in the Solar System—up to 500 parts per million. This is, however, still much lower than the water in the pre- or protosolar nebula, and in the interstellar gas at large. This is probably because long period comets formed near the orbits of Jupiter, and that planet's gravitational influence hurled them out to the Oort Cloud. Nevertheless, comets contain material largely untouched by chemical and physical processes—pristine material from the earliest times of our Solar System. This is one reason why space scientists have been so keen to send space probes to comets—including NASA's Stardust mission, which brought back samples of dust from comet Wild 2, and the European Space Agency's Rosetta mission, which landed a probe, Philae, on the surface of comet 67P/Churyumov–Gerasimenko.

Ice Giants

At the inner edge of the Kuiper Belt, at 30 AU, is the orbit of Neptune. Its size and composition are very similar to that of Uranus, which orbits at around 19 AU. These two planets are *ice giants*. The word *ice* here refers not just to water ice but also to solids of other volatile compounds, primarily methane and ammonia—but water makes up around 65 percent of the mass of both Uranus and Neptune. The measured D/H values for water on Uranus and Neptune are about 45 parts per million—lower than one might expect for objects formed outside the snow line, and only about one-tenth that of Oort Cloud comets. There are various hypotheses for why this should be so. The most promising is that much of the water these planets hold was actually made from the reaction of hydrogen gas (with a very low D/H ratio) with carbon monoxide.[9] Another idea is that these planets formed much closer to the snow line and were shunted further out by collisions or gravitational interactions with other objects. Finding out more about these two planets is a major goal of NASA within the time frame 2024–2037—not least because "Ice giants appear to be very common in our galaxy" and they "challenge our understanding of planetary formation, evolution, and physics."[10]

The ice giants' water is not ice as we know it. Instead, it is likely a mixture of water beyond its *critical point* (more about that in chapter 4)—and *superionic ice*. Four times as dense as ordinary ice, and black in color, superionic ice is a very good conductor of electricity—as good as a metal. Physicists had predicted the existence of this exotic form of ice back in the 1980s, and in 2019 scientists working at the Laboratory for Laser Energetics in Brighton, New York, managed to make some.[11] The scientists blasted a drop of liquid water with X-rays from six of the world's most powerful lasers—creating pressures more than a million times greater than atmospheric pressure, and temperatures of about 4,000°F, inside the water. Within a few billionths of a second, the water's molecules rearranged into superionic ice—a fact confirmed by observing the way X-rays bounced off the atoms inside (a technique called X-ray diffraction). Inside superionic ice, the oxygen and hydrogen atoms exist as O^{2-} and H^{+} ions. The oxygen ions are tightly packed in a cubic arrangement (hence the high density)—with one atom at each corner of each cube and one in the center of each side. The hydrogen ions are *delocalized*, not belonging to any oxygen in particular—in the same way as electrons are delocalized in metals. And so, just as the freely moving electrons in a metal give rise to its conductivity, the same is true for the hydrogen ions inside superionic ice—but in the ice, the charge carriers are protons (hydrogen ions) rather than electrons. It is as if the oxygen ions are a crystalline solid, while the hydrogen ions are liquid weaving among them.

No one yet knows how widespread superionic ice might be in the Universe at large, but the scientists who created it suggest it might at least be the most abundant form of water in the Solar System. Its existence on Uranus and Neptune may provide an answer to a puzzle about the magnetic fields of those planets, which are way off axis. Earth, Jupiter and Saturn have their

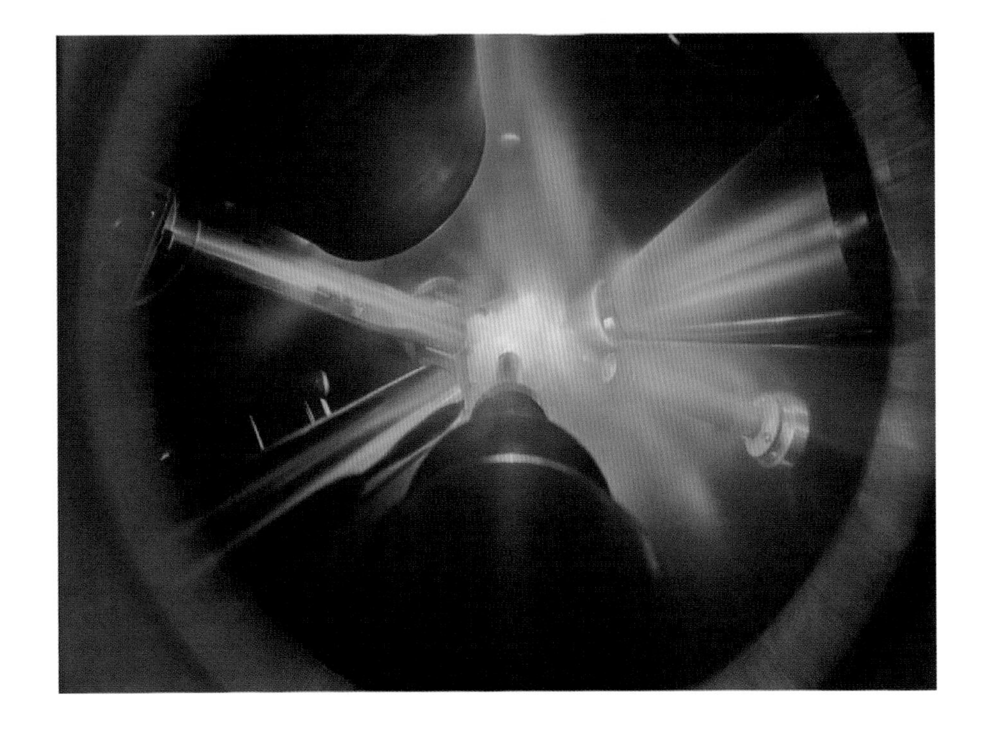

This time-integrated image shows the lasers firing at the water droplet in the experiment that created superionic ice, which was carried out at the Laboratory for Laser Energetics at the University of Rochester. *Source:* Millot, Coppari, Hamel (Lawrence Livermore National Laboratory). Image by L. Krauss (Lawrence Livermore National Laboratory).

magnetic fields aligned closely with their north-south rotational axes. Those magnetic fields are produced by electric currents in a liquid metal core (or in the case of Jupiter and Saturn, liquid metallic hydrogen). Simulations of the magnetic fields of Uranus and Neptune had previously suggested that they are produced by electric currents much further from the core. According to the scientists who fabricated this strange form of water, a thick layer of conductive superionic ice fits the bill perfectly.

Both Uranus and Neptune have ring systems and several moons. The rings are very dim, and made of icy dust grains, each around a twenty-thousandth of an inch in diameter. The moons appear to have a composition just like comets: dark and icy. The largest and best studied is Neptune's largest moon, Triton (which accounts for more than 99 percent of the mass of all Neptune's moons put together). Triton is probably a minor planet from the Kuiper Belt captured long ago by Neptune's gravitational influence. It is between one-fifth and one-quarter ice, mostly in the crust and mantle. Triton's surface shows signs of volcanism—but the volcanoes on Triton do not spew the rocky lava of terrestrial volcanoes. This is cryovolcanism, with a freezing watery mixture emerging onto the surface. This volcanic activity produces plumes of icy material, probably driven by explosions of nitrogen gas beneath the surface.

Icy Triton, with its parent planet Neptune in the background. Neptune's wispy, icy cirrus clouds are made of crystals of solid methane, rather than water ice. *Source:* NASA/JPL.

The Gas Giants

While both of the ice giants in the Solar System are about two-thirds water, the proportion of H_2O in the gas giants, Jupiter and Saturn, is much lower (though in each case, it still amounts to many times the store of water on Earth). They are made nearly entirely of hydrogen and helium, and have long been considered very dry indeed. However, powerful thunderstorms in Jupiter's atmosphere suggest that the proportion of water in the Jovian atmosphere might be greater than previously thought—and in 2018 space scientists found large amounts of water in Jupiter's most famous feature, the Great Red Spot.[12] The water will have

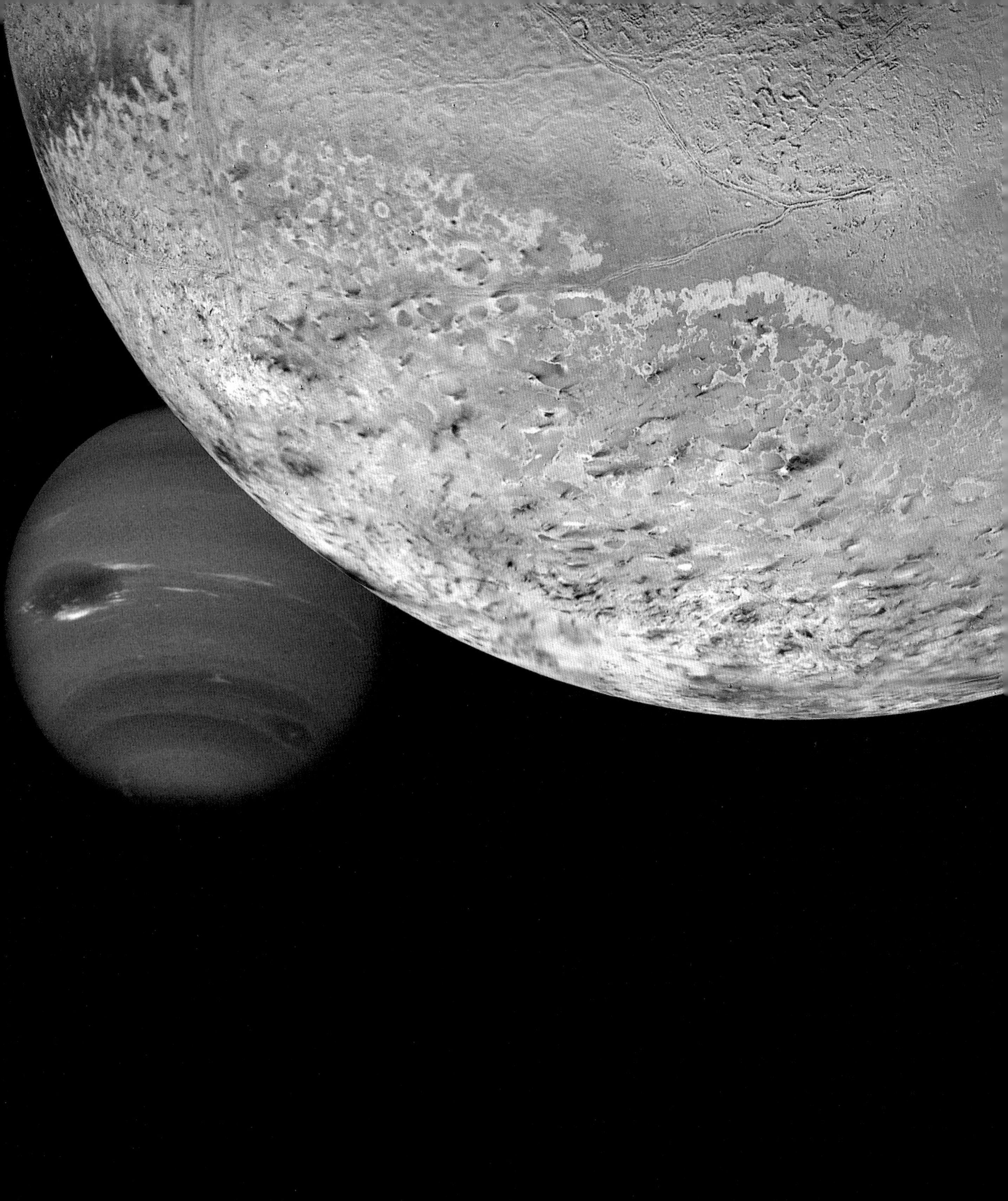

been carried to this storm's cloud tops by convective upwelling—from a more-watery-than-believed atmosphere below.

Even though recent discoveries suggest Jupiter and Saturn may hold more water than previously believed, it is the moons of these two giant planets—and the rings of Saturn—where water abounds. Saturn's rings are made of chunks that are composed mostly of ice, the many billions of pieces ranging from a fraction of an inch to many feet in diameter. There are several competing theories about the origins and the age of the rings—they may be the remains of a moon that broke apart or simply leftover material from the protoplanetary disk of the solar nebula, for example. There is one ring, however, whose origin is known. The extremely wide and diffuse E ring is made of much smaller particles than the others—much smaller than the particles of fine flour. These

particles come from Saturn's moon Enceladus, which, like Triton, is a cryovolcanic world. Jets emerge through fissures on the surface, throwing out material that includes more than 500 pounds of water every second. The source of this material is an ocean of liquid water about 6 miles deep underneath a 20-mile-thick icy surface. The ocean remains liquid as it is heated by volcanic activity in the rocky-icy core. Several of Saturn's other moons orbit within the wide E ring, and their surfaces constantly pick up the icy mixture Enceladus hurls out—and about 5 percent of it even makes it down to Saturn's upper atmosphere. Interestingly, the D/H ratio in water on Enceladus is much higher than in the water of Saturn itself. In fact, the ratio is the same as that in Oort Cloud comets, suggesting that this moon began life as a captured comet.

Between them, Jupiter and Saturn have more than 160 known moons, and, like Enceladus, they all seem to be rich in water. Along with Enceladus, two more of Saturn's moons—Titan and Dione—and two of Jupiter's moons—Europa and Ganymede—stand out. These distant worlds also likely harbor liquid water. Jupiter's moon Europa is one of the most studied and is of the most interest to astrobiologists, for this moon could sustain some kind of life. Jupiter's gravitational field creates forces that stretch Europa, generating heat inside it. Astrophysicists suggest that this heat created and maintains an ocean of liquid water, with an estimated depth of between 50 and 100 miles, below the thick icy surface. Measurements of the way the moon disrupts Jupiter's magnetic field seem to confirm it. Europa's surface is almost completely devoid of craters, probably the result of warmed water seeping out through the surface, just as molten rock does here on Earth, through volcanoes and other fissures in the rocky crust. Dark lines on Europa's surface may be organic compounds similar to those that probably spawned life here on Earth.

In 2015, in the midst of the discoveries of subsurface oceans of liquid water on the moons of Jupiter and Saturn, Jim Green, director of planetary science at NASA headquarters, remarked in a press conference that the Solar System is "now looking like a pretty soggy place."

Jupiter, which has 2.5 times the mass of all the other planets combined. The water-rich Great Red Spot is clearly visible. *Source:* NASA/JPL-Caltech/SwRI/MSSS/Kevin M. Gill.

▲ **NASA's Cassini** spacecraft captured this incredible photograph from within the shadow of Saturn, in 2013. While the planet is eclipsed, the rings, made almost entirely of water ice, are magnificently backlit. The wide, diffuse outermost ring is the E ring. Our watery planet is a blue dot above the E ring below and to the right of Saturn. *Source:* NASA/JPL-Caltech/SSI.

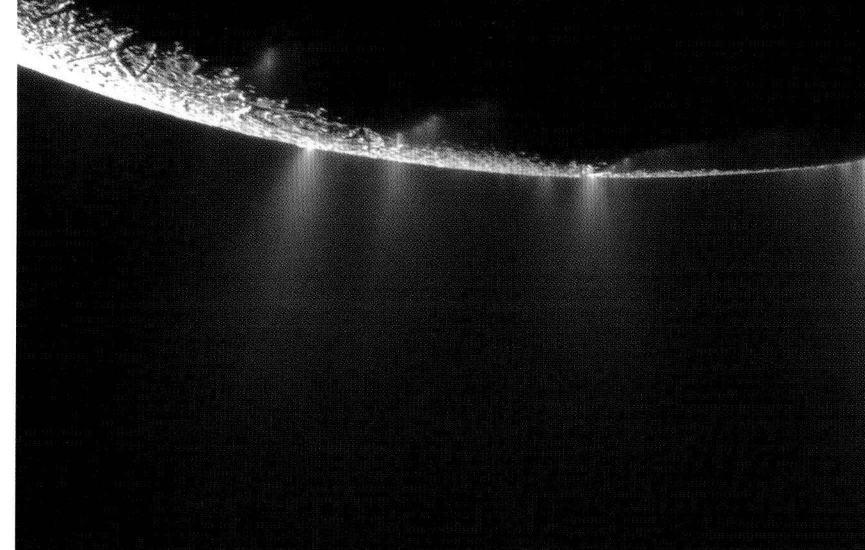

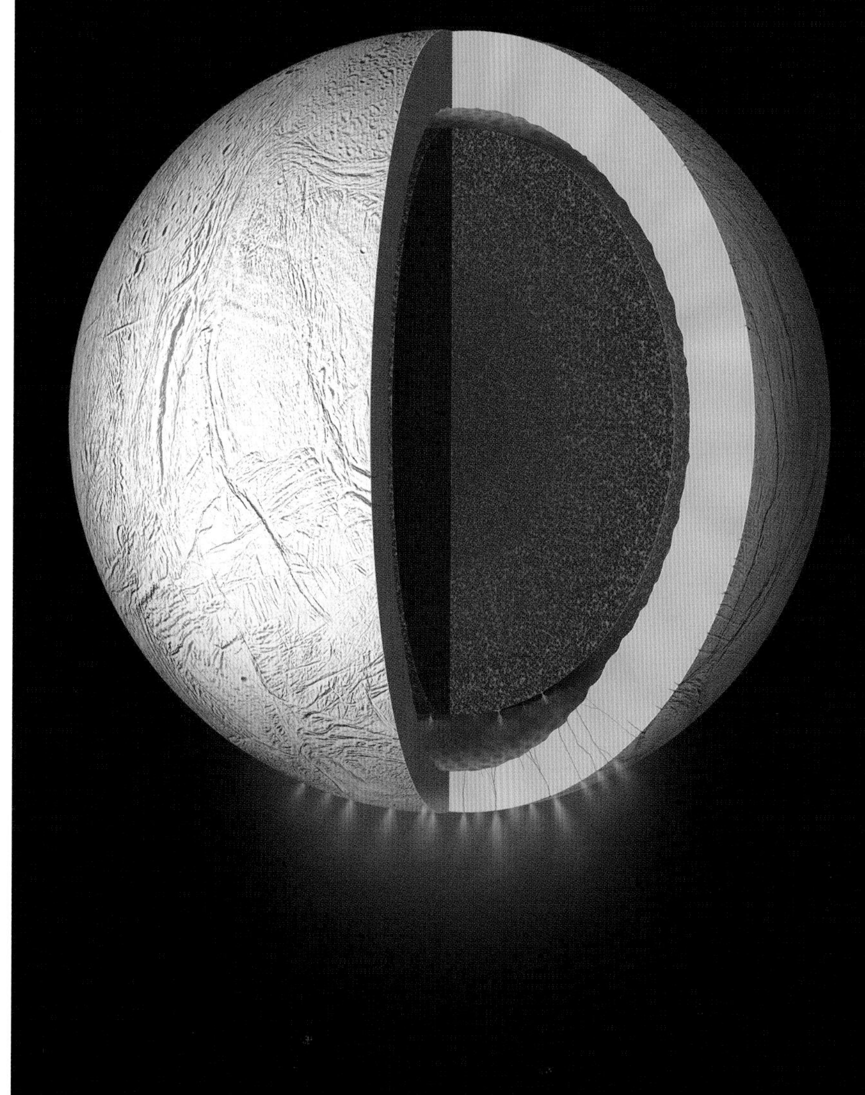

Photograph of Enceladus, backlit by the Sun. The jets of material are more than 90 percent water vapor, which cannot be seen—but the jets are visible because of dust also blown away from the moon's surface. *Source:* NASA/JPL-Caltech.

Artist's rendering of the structure of Enceladus, showing the *source* of the water-rich jets. *Source:* NASA/JPL-Caltech/SSI.

Jupiter's moon Europa, its smooth surface criss-crossed by dark striations, whose nature is not known. One possibility is that these lines are made of organic (carbon-containing) compounds created by reactions in the warm water below the surface. All of this together presents the tantalizing possibility that Europa may harbor some form of life in its oceans. *Source:* NASA/JPL.

Closeup of Jupiter's moon Europa. *Source:* NASA/JPL-Caltech/SETI Institute.

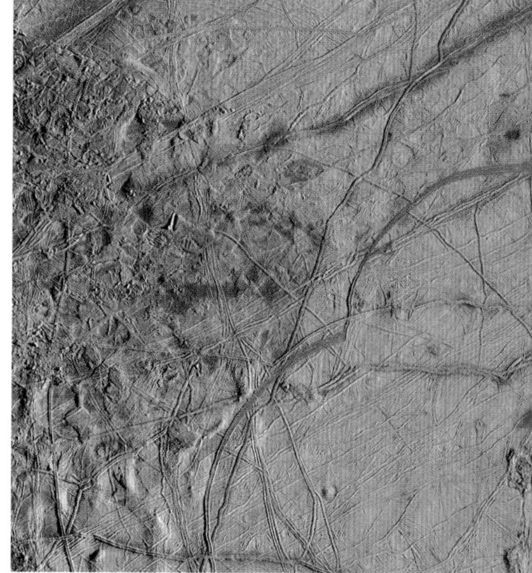

Saturn's moon Dione, which is almost two-thirds water, has a liquid ocean due to cryovolcanism, just like Enceladus. NASA's Cassini spacecraft captured it with Saturn's rings almost perfectly side-on, and Saturn's diffuse atmosphere as a backdrop. *Source:* NASA.

Artist's impressions of auroras on the Solar System's largest moon, Ganymede, with its parent planet Jupiter in the background. In 2015, the Hubble Space Telescope captured images of the auroras, whose locations prove the existence of a subsurface ocean of liquid water about 60 miles deep—an ocean that probably holds more water than all of Earth's rivers, lakes, oceans and ice caps combined. *Source:* NASA.

Inside the Snow Line

Jupiter and Saturn—and, of course, Uranus and Neptune—lie outside the snow line, so it is no surprise that there is plenty of water to be found there. Inside the snow line, which is currently at around 5 AU from the Sun, any ice in direct sunlight will sublime (change directly from solid to vapor), and ultraviolet radiation from the Sun can dissociate molecules in water vapor into hydrogen and oxygen, just as it does to the water released by comets. Nevertheless, water does exist on planets and their moons, as well as asteroids, inside the snow line—including, of course, our own planet.

The Sun was cooler when the Solar System formed than it is today, and would have been partly obscured by material from the Solar Nebula. As a result, the snow line would have been closer to the Sun. Planetary scientists estimate that back then, it would have been at about 2.7 AU.[13] The outermost asteroids, which orbit at more than about 3 AU, consist of up to 20 percent water to this day. Most of these are C-type asteroids, which consist mostly of clay and silicate minerals. These minerals are hydrated; that is, they have water molecules locked away in their crystal structures. As well as being trapped in hydrated minerals, water is often present in tiny spaces called "fluid inclusions" within the structure of the rock.

Studies show that C-type asteroids are very similar to a class of meteorite called *carbonaceous chondrites*, which in turn are very similar to the composition of the Sun (minus the hydrogen and helium). These meteorites, then, are like leftover pieces of the Solar Nebula. The *chondrite* part of the name refers to chondrules, which are glassy spheres of rock solidified in the early Solar System, before accreting to form asteroids. The largest of the asteroids—so large, in fact, that it was reclassified as a *dwarf planet* in 2006—orbits at the cusp of the early Solar System's snow line, at about 2.8 AU. Ceres, as it is called, has an estimated water content of 20 percent, mostly as ice that fills pores in the rocky crust and in hydrated minerals on the surface. It even has a 2.5-mile-high cryovolcano that produces an outpouring of a mixture of clay and salty, slushy ice.[14] The innermost, S-type

The cone of the cryovolcano Ahuna Mons, on watery dwarf planet Ceres, captured by NASA's Dawn spacecraft. *Source:* NASA/JPL/Dawn mission.

asteroids have orbits as close as 2.2 AU to the Sun. They are much drier—although a 2018 study of particles collected from S-type asteroid Itokawa, by the Japan Aerospace Exploration Agency's Hayabusa spacecraft, found water content of about 1 percent.[15]

Mercury and the Moon

If the S-type asteroids, orbiting more than 2 AU from the Sun, contain so little water, then Mercury should be bone dry. Planet Mercury orbits an average of just 0.4 AU from the Sun. Not only do surface temperatures reach as high as 800°F; the solar wind blows more strongly there than on any other body in the Solar System, blasting away any water that might otherwise persist. And yet there is water on Mercury—though not a great deal of it.

Nearly all of Mercury's water is ice that lurks in *permanently shadowed regions*: deep craters around the planet's poles, which are never exposed to the Sun. Earth-based observations hinted at this possibility for decades, and it was confirmed in 2015 by NASA's MESSENGER orbiter. The source of that ice would probably have been impacts with comets and asteroids early in the

Solar System's history. Trace amounts of water vapor have also been detected in Mercury's thin atmosphere, probably originating from ice in the polar craters that has sublimed—although there is another probable source of at least some of Mercury's water. Energetic protons (nuclei of hydrogen-1) and deuterons (nuclei of hydrogen-2, made of a proton plus a neutron), thrust onto the planet by the solar wind, react with oxygen atoms in the planet's rocks. This same reaction is likely to happen on the Moon. Several teams of researchers have successfully mimicked the process here on Earth, by bombarding minerals similar to mercurial and lunar soils with beams of hydrogen ions under high vacuum.[16]

The Moon is our nearest neighbor and has been studied in much more depth than Mercury—people have even visited, and brought back samples of rock and soil. Some water was detected in samples of lunar soil (known as *regolith*) that Apollo astronauts brought back to Earth in the 1970s—but it was deemed likely that the samples had been contaminated with terrestrial water vapor. Still, lunar scientists since the early 1960s had proposed that water might exist in deep craters at the lunar poles, or even trapped in the regolith more widely. A few years after the end of the Apollo missions, in 1976, the un-crewed Soviet probe Luna 24 brought back to Earth samples it had collected from various depths within the regolith—down as far as 6 feet. Water made up 0.1 percent of those samples, proving conclusively that water is indeed present on the Moon. A 2008 reevaluation of the Apollo samples—specifically a study of the inside of glassy beads that could not have been contaminated—led to the conclusion that water really was present in those samples, too.

Several missions have searched for stores of water ice in the Moon's craters. In 1994, the US military probe Clementine bounced radio waves off the floor of the deep craters at the Moon's south pole. The echoes of those radio waves, detected back on Earth, suggested that large quantities of water ice were indeed present—but results were inconclusive. NASA's Lunar Prospector orbiter, which arrived at the Moon in 1998, carried a neutron detector. The detector collected neutrons that had been delivered to the Moon in the solar wind and had interacted with hydrogen atoms in water molecules en route. Based on analysis of the energies of those neutrons, the mission's scientists estimated that more than 3 billion tons of water might exist in the polar craters. At the end of its mission, in 1999, engineers directed Lunar Prospector to use its last reserve of fuel to change course, so that it would crash into the crater Shoemaker, near the lunar south pole. The resulting plume of debris, studied from Earth, did not reveal the presence of water. Many reasons were put forward for that negative result, including the possibility that the spacecraft may have just missed its target.

The Shackleton crater, 13 miles wide and 2.6 miles deep, lies directly at the Moon's south pole. Its steep sides are illuminated by sunlight. This means that it could one day house solar panels that could produce electricity to split water in the crater into hydrogen and oxygen—perfect for making rocket fuel. *Source:* NASA's Scientific Visualization Studio.

Interest in water in lunar craters gathered more pace in the 2000s. In 2008, Japan's space agencies sent SELENE, better known as Kuguya. One of its aims was to produce images of the polar craters' floors, to find evidence of water ice—but such evidence was not forthcoming. Like Lunar Prospector, Kaguya was deliberately crashed into a crater, again with inconclusive results. More definitive results did eventually come: there really is a lot of water on the Moon, it seems. First, the Indian orbiter Chandrayaan-1 sent a probe crashing into the Shackleton crater in the lunar south pole. On its way down, the probe detected water vapor in the thin lunar atmosphere, while a NASA instrument called the Moon Mineralogy Mapper onboard the orbiter detected possible water ice on the surface. The results suggested that water is constantly being created, retained and vaporized, in a kind of tenuous lunar water cycle. Some of the vapor finds its way to the *cold traps* provided by craters in polar regions, where it can stay for longer periods. Overall, the results suggest

Analysis of data from NASA's Moon Mineralogy Mapper led to these maps of the distribution of ice (blue) around the Moon's south pole (left) and north pole (right). Gray colors have been added to represent surface temperatures—the darker the gray, the lower the temperature. *Source:* NASA/JPL.

that ice exists mostly in small chunks mixed in with the regolith, rather than extensive sheets. This notion was given further weight by yet another planned crash into a crater, this time by the second stage of the rocket that carried NASA's Lunar Crater Observation and Sensing Satellite (LCROSS) to the Moon. LCROSS flew directly into and through the debris ejected by the collision. Onboard instruments found plenty of hydroxyl radicals (*OH)—a reactive molecule that probably would have come from water ice or hydrated minerals. In 2018, a fresh look at the data gathered by the Moon Mineralogy Mapper discovered definitive evidence that large amounts of water exist at the lunar poles—not only in the always-dark craters, but also mixed in with the regolith.[17]

Mars and Venus

If there is water on Mercury and the Moon—small, airless and exposed, and well within the snow line—then Mars and Venus must have plenty. In the distant past, these two worlds almost certainly had plenty of water, perhaps as much as Earth does now—but unlike our planet, they have lost most of it.

Mars certainly is a dry world, but a series of exciting discoveries over the past few decades have shown that there is at least

The European Space Agency's Mars Express orbiter captured this image of swirls of water ice and dry ice at Mars' north pole. *Source:* ESA/DLR/FU Berlin; NASA MGS MOLA.

more water than previously believed. Since the planet's average surface temperature is a chilly −80°F (and much colder near the poles), it is little surprise that nearly all of the water on Mars is in the form of ice. The planet has two permanent polar caps, which are made primarily of water ice. The northern polar cap contains much more permanent water ice, as well as large amounts of sand. Both polar caps have a swirled appearance, created when spiraling winds caused by the planet's rotation excavate the frost. When winter comes to the north pole, a layer of frost made of water ice and dry ice (frozen carbon dioxide) forms on top of the water ice. At the south pole, the water ice is covered by dry ice all year round. Around Mars' polar caps are several craters that act as cold traps, although they collect a lot more water ice than the cold trap craters on the Moon and Mercury. The most spectacular is the crater Korolev, just over 50 miles in diameter.

In addition to the visible deposits of ice, there is plenty of water ice beneath the surface. For example, in 2019, measurements by

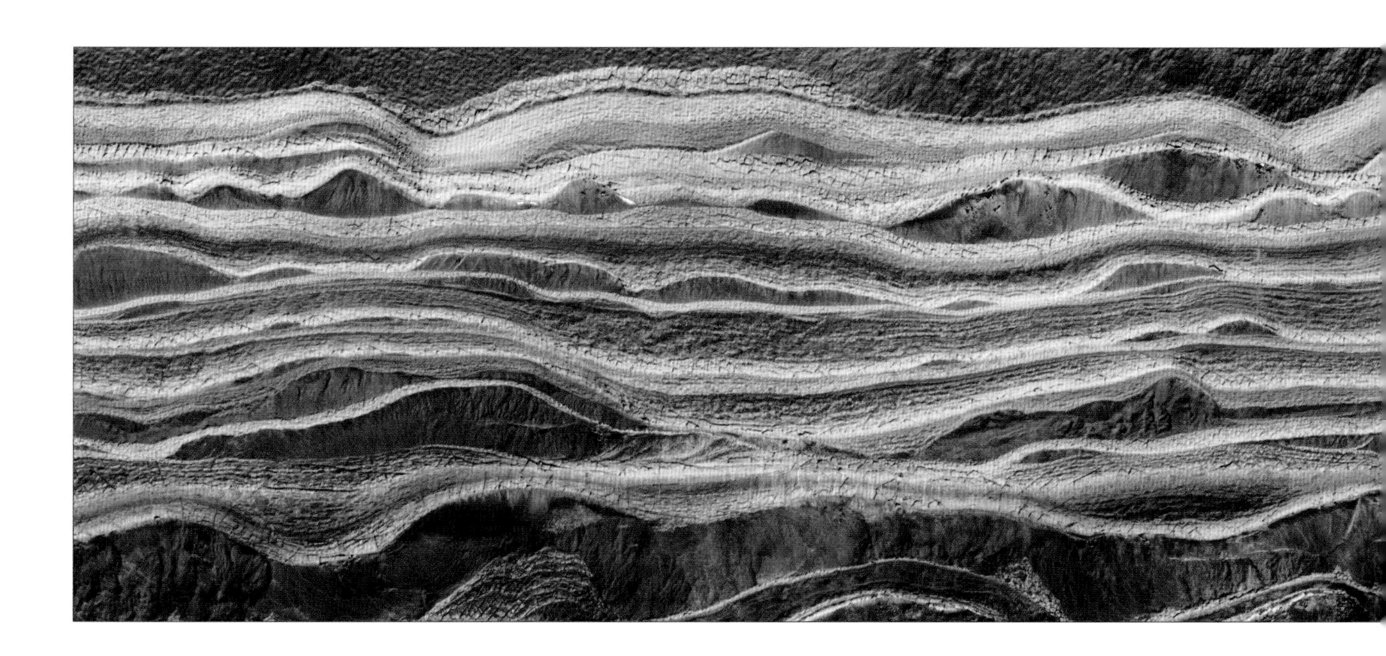

This remarkable picture of Mars' Korolev crater is constructed from images captured during five orbits by the High Resolution Stereo Camera onboard the European Space Agency's Mars Express spacecraft. The crater has a permanent store of about 2,200 cubic miles of water ice. *Source:* ESA/DLR/FU Berlin, CC BY-SA 3.0 IGO.

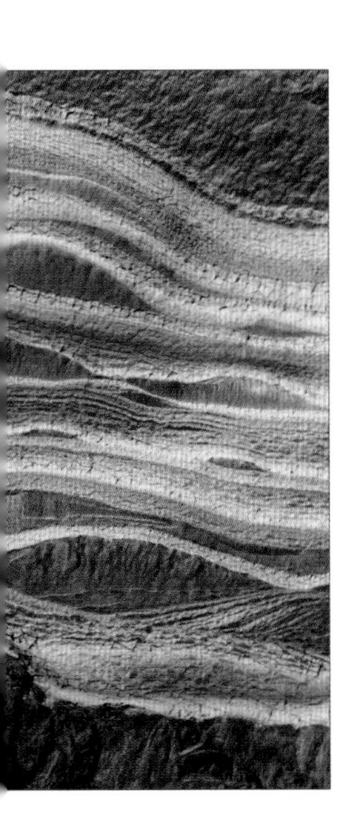

Color-adjusted Mars Reconnaissance Orbiter image of exposed regions around Mars' north pole. The white lines are water ice, the blue layers sand. *Source:* NASA/JPL/University of Arizona.

instruments onboard NASA's Mars Reconnaissance Orbiter led to the discovery of layers of ice a mile beneath the ice cap at the north pole[18]—remnants of a more extensive ice cap from Martian ice ages long ago. At the end of each ice age, some ice remained, and was covered in sand—so the layers constitute a record of climate change on Mars. In places around the ice caps, this stratification has become exposed. Also in 2019, a team of scientists reviewed data collected by a range of instruments carried on several Mars probes, in an attempt to discover how deep this ice lies in the Martian regolith.[19] They found huge reserves of ice just a few inches beneath loose soil across large areas—making it easy enough for future crewed missions to dig it up by hand. Altogether, about 1.2 million cubic miles of ice have been identified at or near the surface of Mars. If this were melted, it would cover the whole planet to a depth of more than 100 feet.

Perhaps most exciting of all the water found on Mars so far is the discovery of liquid water. In 2018, Italian scientists using data from ground-penetrating radar onboard the European Space Agency's Mars Express Orbiter discovered a permanent lake of liquid water beneath the Martian surface.[20] The lake is about 12 miles long and a mile beneath the surface, underneath the layers of sand and ice, about 300 miles from the planet's north pole. The temperature below the surface is likely higher than at the surface, but still well below freezing; the water probably remains liquid thanks to the antifreeze effect of a high concentration of dissolved salts. The scientists compared this body of water with the subglacial Lake Vostok, in Antarctica—and since *extremophile* organisms (living things that thrive in hostile conditions) have been found living in Lake Vostok, perhaps some form of life might one day be discovered in the lake under the surface of Mars, too.

There are many features visible on Mars today, including dried-up river courses and lake beds, that give clues to that warmer, wetter past. In 2017, NASA's Curiosity rover found what look like cracks in ancient dried mud, and in 2019, concentrated deposits of mineral salts that give testimony of shallow lakes that repeatedly evaporated and refilled.[21] Also in 2019, a team of scientists found compelling evidence of a vast ancient system of groundwater.[22]

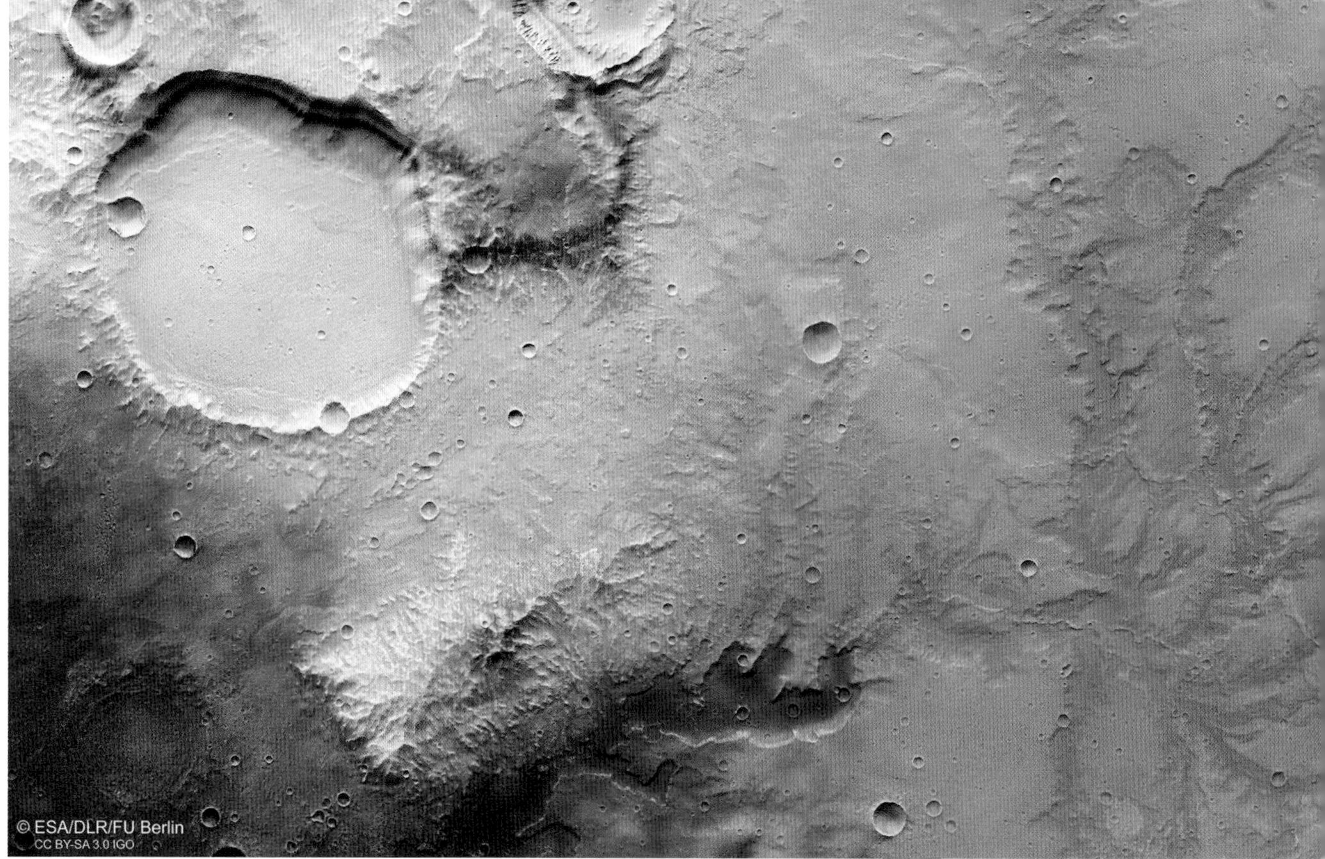
© ESA/DLR/FU Berlin
CC BY-SA 3.0 IGO

Artist's model of the huge ocean that may have existed on Mars billions of years ago, created by virtually filling the planet's landscape with water and adding cloud cover. The model used data from NASA's Mars Atmosphere and Volatile Evolution (MAVEN) mission. *Source:* NASA/MAVEN/Lunar and Planetary Institute.

The scientists used the European Space Agency's Mars Express orbiter to study 24 craters deep enough that they would have been below the proposed ancient sea level. They found evidence of the upwelling and retreat of water, again reflecting rapid changes in the Martian climate.

So what happened to that plentiful supply of water that Mars once had? The short answer is that most of the water evaporated to the upper atmosphere, from where it was lost to space—blown away by the solar wind and dissociated into hydrogen and oxygen by ultraviolet radiation. The hydrogen atoms released by dissociation are very light and much more prone to escape than the oxygen atoms. And of the two isotopes, hydrogen-1 (ordinary hydrogen) is much more likely to escape than hydrogen-2 (deuterium), because a deuterium atom (D) has twice the mass of a regular hydrogen atom (H). For a given temperature, lighter atoms

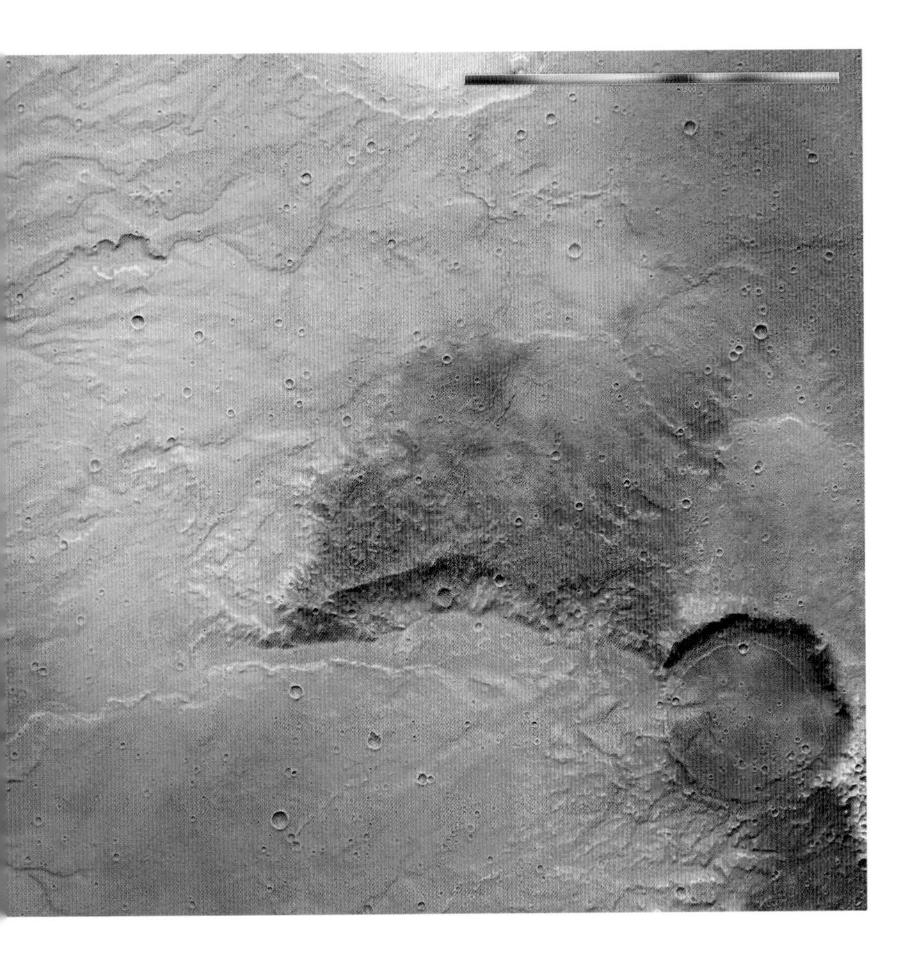

High resolution color-coded image captured by the High Resolution Stereo Camera on the European Space Agency's Mars Express spacecraft. The image shows a dried-out river valley on Mars; red represents the highest altitude, blues and purples the lowest. *Source:* ESA/DLR/FU Berlin, CC BY-SA 3.0 IGO.

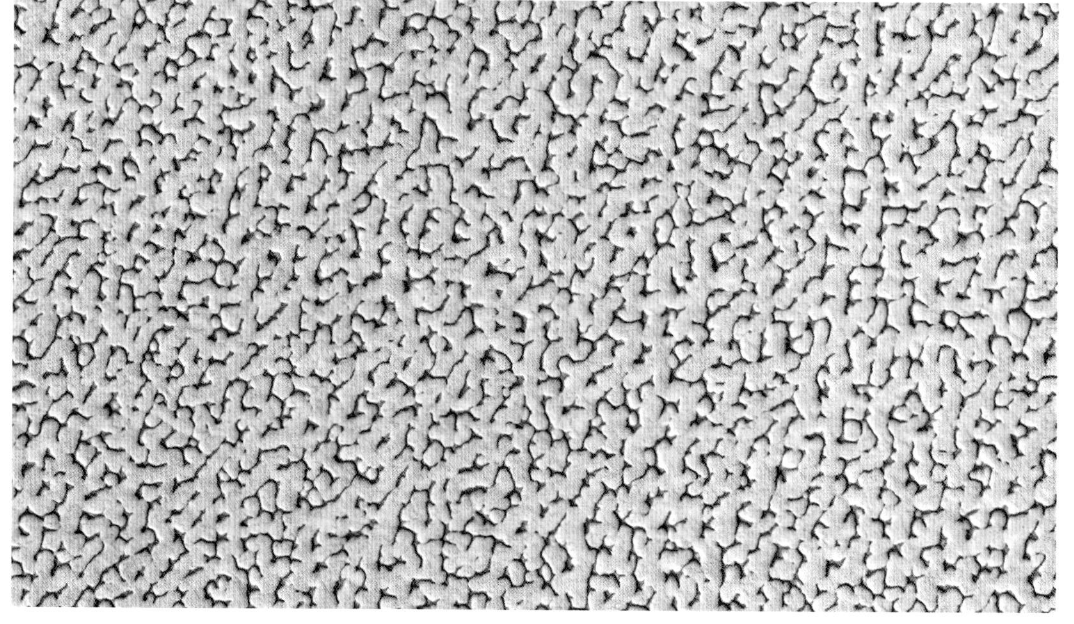

have a higher average speed than heavier ones. As a result of the preferential loss of ordinary hydrogen atoms from the top of the atmosphere, the relative proportion of deuterium atoms—and therefore the D/H ratio—in the water in Mars' atmosphere today is much higher than it was. The change in the D/H ratio can give clues to just how much water Mars has lost. A 2016 study that did just that estimated that that around 3.5 billion years ago, Mars probably had a vast sea that held more water than the Arctic Ocean here on Earth—enough water to cover its entire surface in a liquid layer about 450 feet deep.[23]

Mars once had a strong magnetic field, created by electric currents in its molten core spinning around a solid core—the same mechanism that creates Earth's magnetic field. A strong magnetic field protects a planet from the solar wind, diverting energetic charged particles around the planet. Mars is much smaller than Earth, so its core solidified long ago—and that led to a significant weakening of its magnetic field around 3.5 billion years ago. Around the same time, the Sun's output began to increase, making Mars even more vulnerable. Mars' small size also means it has less ability to hold on to its atmosphere, so it could not benefit from the greenhouse effect—which, here on Earth, keeps our planet's surface nearly 60°F warmer than it would otherwise be. On top of all of this, variations in the planet's orbit gave rise to warmer periods: water evaporated into the upper atmosphere, making it more susceptible to ultraviolet radiation.

The Curiosity rover snapped this image of a slab of dried mud, about 4 feet wide in Gale Crater, in 2017. The cracks provide evidence of rapid drying and are an estimated 3.5 billion years old. *Source:* NASA/JPL-Caltech/MSSS.

The High Resolution Imaging Science Experiment onboard the Mars Reconnaissance Orbiter captured this intriguing image during spring at Mars' north pole. The white color is actually dry ice (solid carbon dioxide), which is subliming to reveal water ice below as the surface warms. *Source:* NASA/JPL-Caltech/Univ. of Arizona.

Like Mars, Venus was almost certainly much wetter in the past, though evidence is more scant. Very few missions have made it to the surface, because of the extremely hostile conditions there and the thick clouds of sulfuric acid that completely obscure surface features from view. But while light cannot penetrate those clouds, infrared radiation can. In 2009, the European Space Agency's Venus Express Orbiter detected infrared radiation emitted by the planet and found that much of Venus' surface is granite.[24] On Earth, granite is formed when ancient rocks in the seafloor, made of basalt, are forced down toward the mantle,

and melt. The rock takes water with it, which becomes locked in the composition of the minerals that make up the granite rock that forms when the molten rock cools. The same process is likely to have occurred, on a much wetter Venus, long ago. And in 2016, a team of planetary scientists using a supercomputer at the NASA High-End Computing (HEC) Program carried out detailed three-dimensional climate simulations for the ancient Venusian atmosphere. They used topographical data (relating to the planet's physical geography) gleaned from the Magellan mission of the early 1990s, together with knowledge of how bright the Sun would have been at the time. To their simulation, they added a volume of liquid water to form an ocean and an atmosphere, both based on reasonable assumptions and available evidence. When they let the simulation run, they observed a mild climate—a world with rain clouds and rivers that could have sustained life.[25] Today, water vapor makes up just 0.002 percent of Venus' atmosphere—1/200th the concentration in Earth's atmosphere—and the surface temperature averages just over 860°F, hot enough to boil water instantly, even under the extremely high Venusian atmospheric pressure.

▲ **Reprocessed** Mariner 10 image of Venus, showing the thick clouds of sulfuric acid droplets that shroud the planet. *Source:* NASA/JPL-Caltech.

What happened to Venus' water is similar to and different from what happened on Mars. Venus orbits an average of 0.7 AU from the Sun (compared with Earth's 1.0 AU, of course). This means it receives more intense solar radiation than Mars and Earth, with enough power to evaporate huge amounts of water from those early oceans—water vapor would have made up more than one-fifth of the atmosphere. Since water vapor is a very effective greenhouse gas, that would have raised surface temperatures still further—something planetary scientists call a *moist greenhouse effect*. Some of the vapor would have condensed to form rain, but high in the atmosphere, solar ultraviolet radiation would have dissociated water, just as it did on Mars. And just as happened on Mars, the resulting hydrogen and oxygen ions are lost to space. This continues to this day: in 2008, the European Space Agency's Venus Express spacecraft detected a loss of a trillion trillion hydrogen ions every second from Venus' atmosphere. Since ordinary hydrogen is more likely to be lost than heavier deuterium, this loss of hydrogen ions increases the D/H ratio. That ratio for water in Venus' atmosphere is 50,000 parts per million—more than 250 times the D/H ratio for water in Earth's atmosphere (and oceans).

Drops to Drink

Coleridge's inspiration for "The Rime of the Ancient Mariner" was the epic voyages of exploration of the seventeenth and eighteenth centuries—journeys to undiscovered places—during which, unfortunately, the unfamiliar was often mistaken for the uncivilized or an opportunity to plunder mineral riches. The character of the mariner was based on an English sailor, Simon Hatley, who killed an albatross during a privateering expedition. Hatley was part of a private mission with permission from the British Navy to attack Spanish ships and keep their riches. There are parallels in the exploration of space today—though none so brutal, hopefully, and this time with no intelligent life to disturb, as far as we know. Private companies, such as Planetary Resources in the United States and the Asteroid Mining Corporation in the United Kingdom, have the expressed purpose of mining asteroids or the Moon. One of the precious resources such missions will be hunting for is, of course, water. Not only would it provide the missions' astronauts with water to drink; it could also help them get back home. Electric currents generated by vast solar farms could separate water into hydrogen and oxygen, which together make the perfect rocket fuel. This is particularly relevant for future explorers wanting to use a lunar base as a stepping-stone to human exploration further out in the Solar System, so water out there in space could help us in our own epic voyages.

For now, though, the only place where we humans can take advantage of plentiful water is here on Earth. The next chapter examines the source of our planet's water and our long relationship with it.

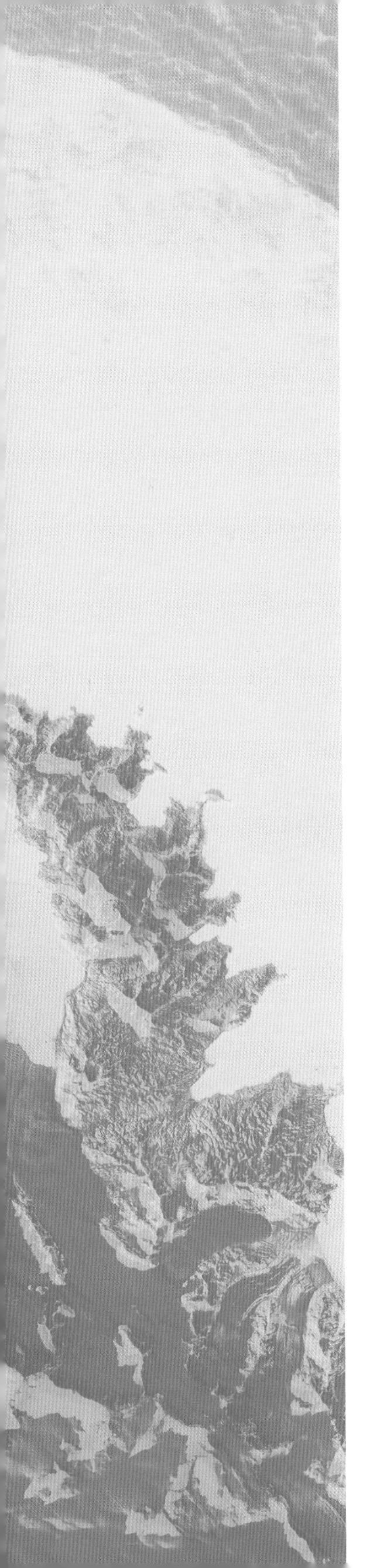

CHAPTER 2

Blue Planet

The great Italian polymath Leonardo da Vinci was fascinated by water. Acknowledging its contradictory nature, he wrote that water is "sometimes health-giving, sometimes poisonous" and that it "hollows out or builds up, tears down or establishes, fills or empties . . . is the cause at times of life or death, of increase or privation, nourishes at times and at others does the contrary." Here on Earth, water—passing endlessly from sea to air to land and back to sea—has indeed done all those things. It has also shaped the land and has enabled and nourished life on our planet. Fresh (non-salty) water makes up only 2.5 percent of Earth's supply, and more than two-thirds of that fresh water is locked away in glaciers and ice sheets—and most of the rest is underground. No wonder, then, that people have fought over water, in local skirmishes and in major wars—and civilizations have fallen as a result of droughts. Leonardo concluded: "In time and with water, everything changes."

Blue light scattered in all directions by air molecules (including up into space) makes the atmosphere appear blue, just as it does when looked at from the ground. But the main contribution to the blue color of our planet is the oceans, which cover 71 percent of the surface. *Source:* NASA Earth Observatory.

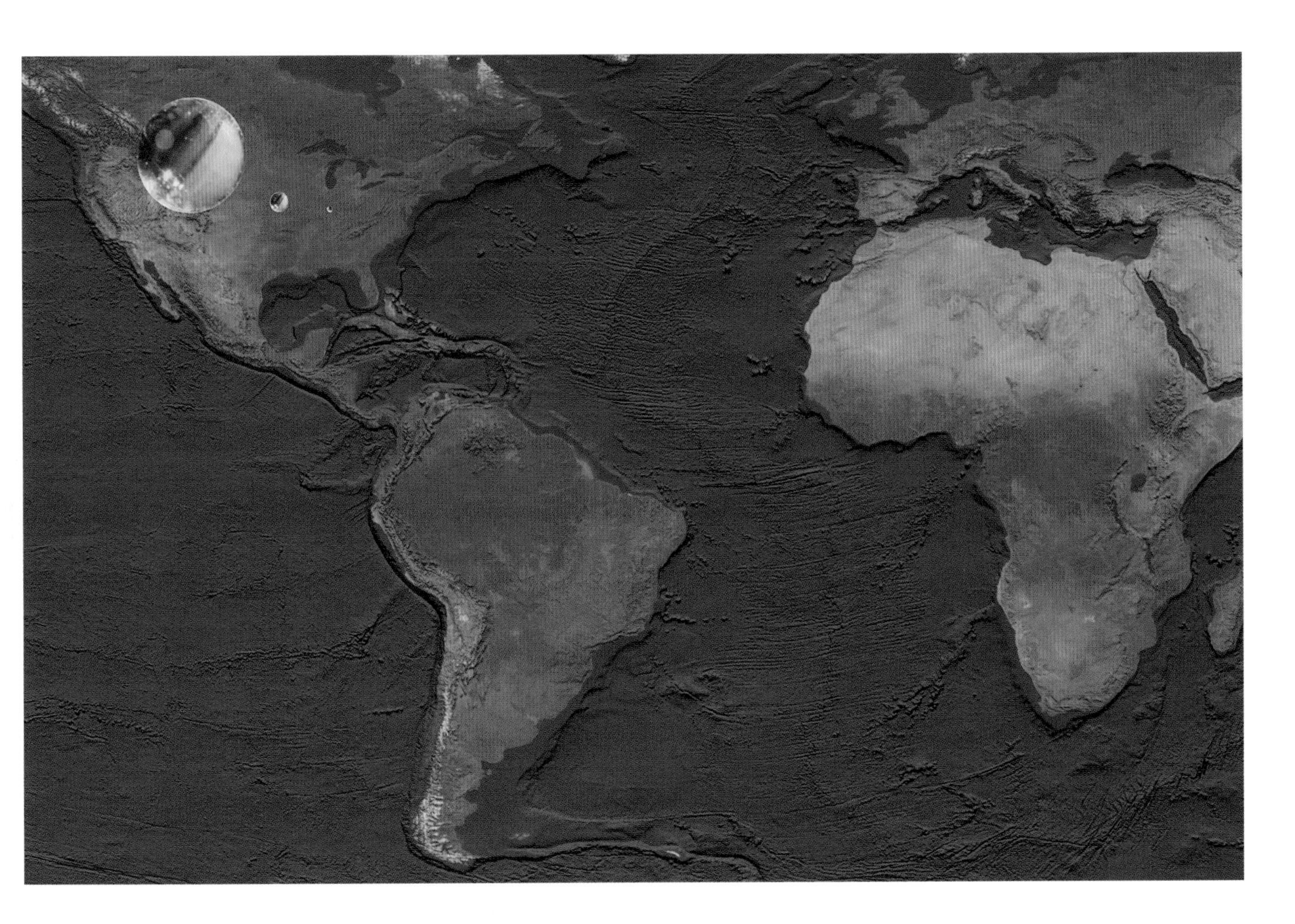

In this image, based on a graphic produced by the US Geological Survey, the total amount of water on Earth is shown as the large drop. Most of that is salty water in oceans and lakes. The total amount of non-salty (fresh) water is represented by the small drop next to it. All the fresh water in rivers and lakes—1/14,000th of the total, is represented by the tiny drop next to that. *Sources:* Background images: NASA/Goddard Space Flight Center Scientific Visualization Studio US Department of Commerce, National Oceanic and Atmospheric Administration, National Geophysical Data Center, 2006, 2-minute Gridded Global Relief Data (ETOPO2v2). http://www.ngdc.noaa.gov/mgg/fliers/06mgg01.html. The Blue Marble Next Generation data is courtesy of Reto Stockli (NASA/GSFC) and NASA's Earth Observatory.

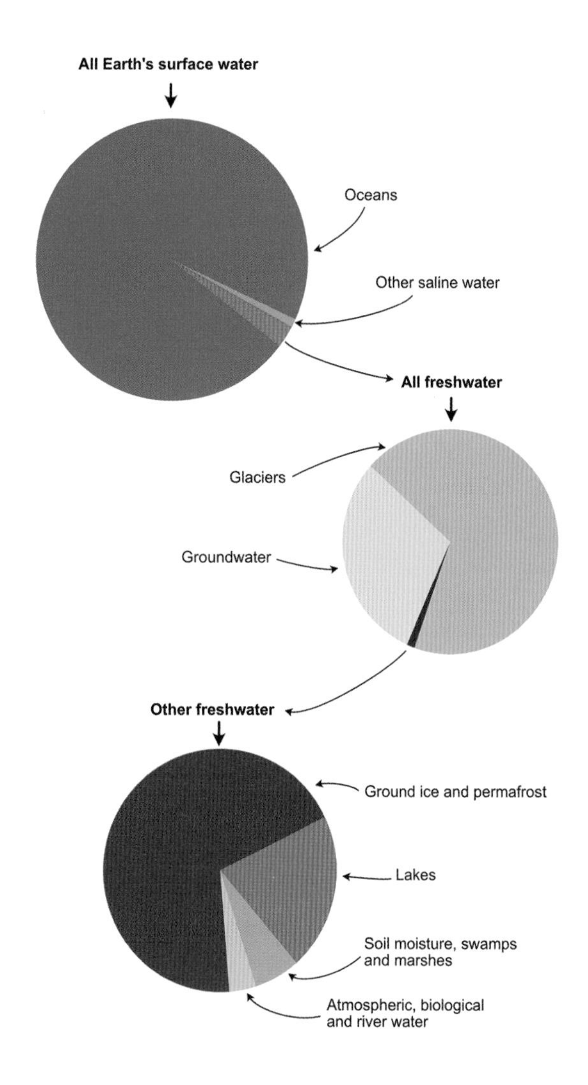

Nearly all of Earth's water is in the oceans. Most of the freshwater is locked away in glaciers. Liquid freshwater at the surface represents a tiny percentage of the whole. *Source:* Adapted from Igor Shiklomanov, "World Fresh Water Resources," in *Water in Crisis: A Guide to the World's Fresh Water Resources*, ed. Peter H. Gleick (New York: Oxford University Press, 1993). Via US Geological Survey.

Where Did Our Water Come From?

Like Mars, planet Earth has had its fair share of dramatic climate changes—but unlike Mars, most of our water is still with us. This is largely due to the fact that our magnetic field is still going strong, pushing back against the solar wind so that it cannot blast away our atmosphere. Thanks to that atmosphere, we have a greenhouse effect, as on Venus, which makes our planet warm enough for a good deal of our water to be liquid. But unlike on Venus, very little of our water has been carried high into the atmosphere, where ultraviolet radiation would break it apart and send the fragments flying into space. We have been saved that fate because our planet has an *atmospheric cold trap*. Like the cold traps in those dark craters on Mercury and the Moon (see chapter 1), an atmospheric cold trap turns water vapor into water ice, which stops it from escaping. Our cold trap is at an altitude of about 10 miles, at the top of the bottommost layer of our atmosphere, the troposphere. The air's temperature decreases rapidly from the ground up, and at the top of the troposphere (the tropopause), it is so cold that ice has to form. That is where wispy cirrus clouds precipitate—and it is why the tallest thunderclouds have flat tops. The clouds eventually precipitate the water back down in the form of rain or snow. By freezing out the water vapor, Earth's cold trap also prevents the atmosphere from becoming too humid, as happened on Venus—so we have thus far avoided the moist greenhouse effect and then the runaway greenhouse effect that befell that planet. (That will, however, happen one day. The Sun will quite naturally become hotter over the next few hundred million years, and planetary scientists estimate that Earth's water will indeed suffer the same fate as the water on Venus—though we should be safe for the next billion years.[1])

And so, for now at least, ours is a world with plenty of water—as solid, liquid and vapor. But where did all that water come from? In short, we still do not know for sure—but there are several hypotheses, and there is plenty of evidence. The hypothesis that has prevailed for many years goes something like the following.

Our planet formed by the accretion of asteroid-sized objects called *planetismals*. Since these planetismals were themselves from inside the snow line, they would have been pretty dry. Most of the water on and in our planet must therefore have been delivered later on, by water-rich impacting objects that originated beyond the snow line. The D/H ratio is useful here to determine what kinds of objects might have brought our water. The D/H ratio of water on Earth's surface is close to the ratio for water in C-type asteroids, as measured in chondritic meteorites that are known to come from such asteroids—but much lower than the figure for comets. So, tempting though it is to think of those dirty snowballs as the bringers of water, it probably was not them.

There is lots of evidence to support the scenario in which asteroids delivered our water. For example, it is likely that within a few tens of millions of years of the birth of the Solar System, a Mars-sized object crashed into Earth—an event that formed our Moon. That might well have melted the planet and vaporized any water then present. The asteroids that brought the water, originating from outside the snow line, would have arrived afterward, adding what planetary scientists call a *late veneer*. The idea of this veneer, delivered by *late heavy bombardment*, fits with an idea called the Grand Tack hypothesis. In this scenario, Jupiter formed at about 3.5 AU, moved inward to 1.5 AU and eventually settled at just over 5 AU—like a sailboat tacking in the wind. Each time it changed its orbit, it sent asteroids flying in all directions. Some from outside the snow line could easily have ended up crashing into Earth. The late veneer hypothesis also neatly explains the presence of organic compounds on the young Earth. Such compounds are essential in providing the ingredients for life—and they are abundant on C-type asteroids.

So is that it? Case closed? Our planet formed dry, and then was doused in asteroidal water in a late veneer, during the late heavy bombardment? Well, there are problems with that hypothesis, not least that the D/H ratio of water at Earth's surface is different from the D/H ratio of water in the mantle and the core. In recent years, several researchers have formulated alternative hypotheses, many of which are based on the proposition that most of our planet's water has been here all along. Since our planet formed inside the snow line, the material from which it was made would have had very little water content—but water would not have been entirely absent. Most of the water would have been locked in hydrated minerals in its rocks, and most would have been released when the rocks melted—as a result of energy released by the decay of radioactive elements and by impacts. The layer of molten rock is called a *magma ocean*, and the release of volatile substances such as water in this way called *outgassing*. The magma ocean is a well-developed concept that is also present in the late veneer theory. Careful consideration of the facts has revealed that Earth may have had enough water at the beginning, and retained enough of it, to explain our watery planet's history without much help from asteroids. One intriguing possibility is that much of the water was produced by the dissociation of organic compounds, which are common in the gas clouds in which stars are born and would therefore have been present throughout the solar nebula, not just outside the snow line. In 2020, experimenters heated organic compounds typically found in nebulas, and found that they produced a mixture of oils and water.[2]

Perhaps the most well-developed of the alternative hypotheses was presented in a 2018 study, by a team of planetary scientists at Arizona State University.[3] In a complex but compelling scenario, they utilized recent ideas about the formation of rocky planets, in which large planetary *embryos* can grow within a million years—much faster than the conventional picture, in which planets grow by accreting small particles. These embryos would have contained a small percentage of water, but this would still have constituted several oceans' worth. (Today, water makes up only about 0.02 percent of the total mass of our planet.) Subsequent melting of the planetary embryos' cores would have led to water reacting with iron, and the resulting hydrogen dissolving in the iron—enough to form between three and six oceans (if it were reunited with oxygen, of course). Impacts of other accreting objects would heat the surface of each embryo, melting the rock and creating oceans of magma. The magma would have outgassed water, released from hydrated minerals—and that, they say, was the source of most of the water at the surface today. The embryos would have coalesced together to form Earth, the impact that created the Moon bringing the final piece. Calculations show that the Moon-forming impact may not have thrown all of Earth's water into space after all. The authors' best guess is that, as a result of all these processes, there is the equivalent of four or five oceans' worth of hydrogen at the core, approximately two oceans' worth of water in the mantle—and, of course, one ocean's worth at the surface (that includes all the ice, fresh water, and seawater).

The scenario described above is based on a wealth of data from decades of study and the latest models of planetary formation. But much remains uncertain, and a lot more evidence—from our own planet and across the Solar System—will be needed before we can have a conclusive answer to the question of the source of our planet's water.

Round and Round

While planetary scientists today grapple with complex questions involving isotope ratios and magma oceans, it is perhaps surprising that just 300 years ago, the keenest minds did not understand the basic facts of something as seemingly simple and obvious as the water cycle.

Of course, even in ancient times, people realized that rain contributes to rivers. And since rivers flow into the sea but the sea never fills up, they realized too that ocean water evaporates and is somehow the source of the clouds and rain. But as far as we know, no one supposed that rain alone could be the source of all the water in rivers. The stumbling block for many was that rivers continue to flow for long periods in the absence of rain. Most philosophers who wrote about water in the landscape imagined a planet-wide underground ocean that replenished the world's rivers. The Greek philosopher Aristotle came close to the truth when he suggested that after falling as rain, water may take a long time to trickle through the ground, so that it would fill the rivers for longer than water would if it simply runs off the land. But even he did not believe for a moment that rainwater could be sole source of the world's rivers.

French engineer Bernard Palissy was the first to propose that precipitation alone is indeed the source of river water, in the 1580s.[4] Palissy rejected the idea that prevailed at the time: that water emerging from springs had traveled from the sea, in underground channels. Unfortunately, few took notice of Palissy's ideas at the time, and the idea of subterranean channels persisted. In the 1670s, Palissy's fellow countryman Pierre Perrault

In his 1665 book *Mundus subterraneus* ("Underground World"), German scholar Athanasius Kircher included illustrations of underground caverns constantly replenished with water, via channels that carried water from the sea. *Source:* Cornell University Library.

tested Palissy's idea quantitatively. Perrault estimated the total rainfall within the catchment area of the river Seine, and measured the outflow of that river into the English Channel—and found that the two figures matched reasonably well. Soon afterward, English scientist Edmund Halley carried out an ingenious experiment, in which he measured the rate of evaporation in a pan of water held at the average temperature of the Mediterranean Sea, and extrapolated his results to show that it was quite possible that all the rainfall in the countries surrounding that sea could have come from evaporation of the seawater. Still the idea of a closed water cycle was not accepted in the mainstream until the early nineteenth century.

We now know that the total amount of water in the air at any moment, if condensed to a liquid, would have a volume of more than 3,000 cubic miles. Every day across the world, about 300 trillion gallons' worth of water is carried into the air, after evaporating from oceans, seas, bays, rivers, streams, puddles, leaves, your skin and a host of other locations. And, of course, the same amount of water falls, as precipitation. That amount of water would fill 450 million Olympic-size swimming pools.

A Closed System?

It would be reasonable to suppose that the water cycle is a closed system—in other words, that the water that cycles from sea to air and back to sea is the same water that has been on Earth for billions of years, exactly the same molecules going round and round. It is certainly true that some of the water molecules in your morning coffee or your daily shower are the same molecules once peed out by a dinosaur or shed as tears by William Shakespeare. But it is not true that there is a constant stock of water molecules on our planet: water molecules are constantly being destroyed, and new ones created.

Photosynthesis is one of the culprits. Globally, green plants and certain kinds of bacteria release more than 14 billion tons of oxygen each year,[5] all of it produced by the splitting of water molecules, via photosynthesis (see chapter 6). In order to produce 14 billion tons of oxygen each year, photosynthesis has to rip apart around 16 billion tons of water molecules—equivalent to about 10 million trillion trillion (10^{31}) water molecules every second.

Of course, water is being produced at an almost identical rate, as oxygen reacts with carbon-based hydrogen-containing molecules in combustion or respiration. Consider a candle: wax is made almost entirely of the elements carbon (C) and hydrogen (H), and when a candle burns, those elements combine with oxygen (O) from the air. The result is carbon dioxide (CO_2) and water (H_2O). A typical candle flame produces about 800 million trillion molecules of water per second (a tiny fraction of an ounce).

Respiration—the set of chemical processes by which most living things obtain the energy they need—involves reacting carbohydrates (containing C, H, and O) with oxygen—and again, one of the products is water. The human body typically produces about 10 fluid ounces of water per day as a result of respiration. Some organisms, including birds on extremely long migratory flights, depend on this *metabolic water* as their supply of fresh water.

Some water does leak out of our water cycle altogether, never to return. Around 100,000 tons of hydrogen is lost to space each year,[6] most of it released from the dissociation of water molecules that make it through that atmospheric cold trap. Water is also disappearing deep beneath our feet—carried down into the mantle, as some of Earth's crustal plates are pushed underneath others, by plate tectonics. As an example, nearly 700,000 tons of water is carried under just one subduction zone (the Mariana Trench) each year.[7] Most of the water molecules carried down into the mantle assume positions in the crystal structure of minerals, forming hydrated versions of those minerals. As a result, metamorphic rocks, created at high temperatures and pressures, often contain several percent water by weight (like the granite on Venus). Deeper down in the mantle, water can be split apart, and its constituent atoms forced into the crystal structure of another kind of mineral, described as *nominally anhydrous*. The most plentiful, and best studied, is ringwoodite, which occurs in the transition zone of the mantle, at around 300 miles below the surface. Under the extreme conditions there, hydrogen ions from dissociated water molecules displace magnesium ions in the ringwoodite's crystal structure,[8] while the oxygen ions left over bond with the displaced magnesium. The result is that ringwoodite is able to hold around 2 percent water (albeit as separate hydrogen and oxygen ions).[9] Unlike the water lost to space, the water lost to the mantle can return: it is involved in the *deep water cycle*. When magma comes to the surface, in volcanoes, water escapes from the hydrated minerals and is outgassed into the atmosphere. The deep water cycle is much slower than (although just as important as) the normal, or *hydrological*, water cycle.

Oceans

The best place to start an investigation of the hydrological water cycle is the ocean (oceans, seas and bays), since it accounts for nearly 97 percent of all the world's water. Oceans cover close to 70 percent of Earth's surface, so they are the source of the great majority of water that evaporates into the air. By the same token, they receive most of the rain and snow that falls back down—and they are also the ultimate "sink" for most of the water running off the land, in rivers. If it were not replenished by precipitation and runoff, the water lost from the oceans by evaporation each year would lower the sea level globally by more than 3 feet.

On average, a water molecule will remain in the ocean for around 3,000 years. This is called the *residence time*, and is worked out by dividing the amount of water present by the rate at which water is lost (or, equivalently, gained). By comparison, the residence time of a water molecule in the atmosphere is just ten days. During its stay in the oceans, a water molecule may travel long distances, at varying depths, thanks to ocean currents. At the surface, ocean currents are driven mostly by winds. Variations in temperature and salinity are also important, since they affect the water's density—and currents therefore flow not only horizontally across the surface but also vertically from the surface to the deepest depths.

The ocean currents transport huge amounts of heat around the planet, because water has a particularly high *heat capacity*—in other words, it takes a lot of energy to raise its temperature. The amount of energy required to increase the temperature of a pound of water by one degree would heat a pound of iron by 10 degrees, for example. And so, seawater takes a long time to heat up—and, once warmed, takes a long time to cool. This fact accounts for sea breezes, which blow onshore until late afternoon, and offshore into the evening. In the morning, the Sun quickly warms the land, and air above the land warms, too, and rises. Cooler air from over the sea moves in to take its place, creating the sea breeze. The opposite happens in late afternoon: the land cools more quickly than the sea, and the breeze blows the other way. The

A snapshot from Perpetual Ocean, a visualization of ocean currents carried out as part of "Estimating the Circulation and Climate of the Ocean, Phase II" (ECCO2), a joint project by the Massachusetts Institute of Technology (MIT) and NASA's Jet Propulsion Laboratory. ECCO2 used MIT's general circulation

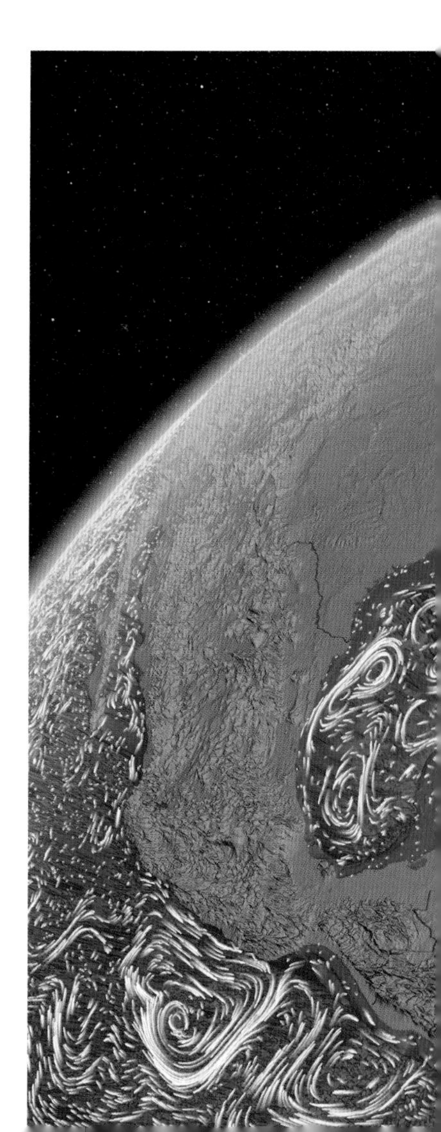

model in conjunction with data gathered in the oceans and by satellite. In this view of surface currents, the Gulf Stream is clearly visible emerging from the Gulf of Mexico, and extending eastward into the North Atlantic Drift. *Source:* NASA/Goddard Space Flight Center Scientific Visualization Studio.

same effect—albeit on a much larger scale both geographically and in its timescale—is the cause of monsoon winds, the best known of which brings dramatic seasonal variations over Southeast Asia. The summer monsoon blows from the Indian Ocean, bringing heavy rainfall; the winter monsoon brings drier air from the heart of the continent down toward the coastal regions.

The high heat capacity of water means that ocean currents play a major role in global climate. One of the major surface currents is the Humboldt Current, which carries cold water up the western edge of South America, from the southern tip of Chile up to the northern coast of Peru. The waters off the coast of Peru can be

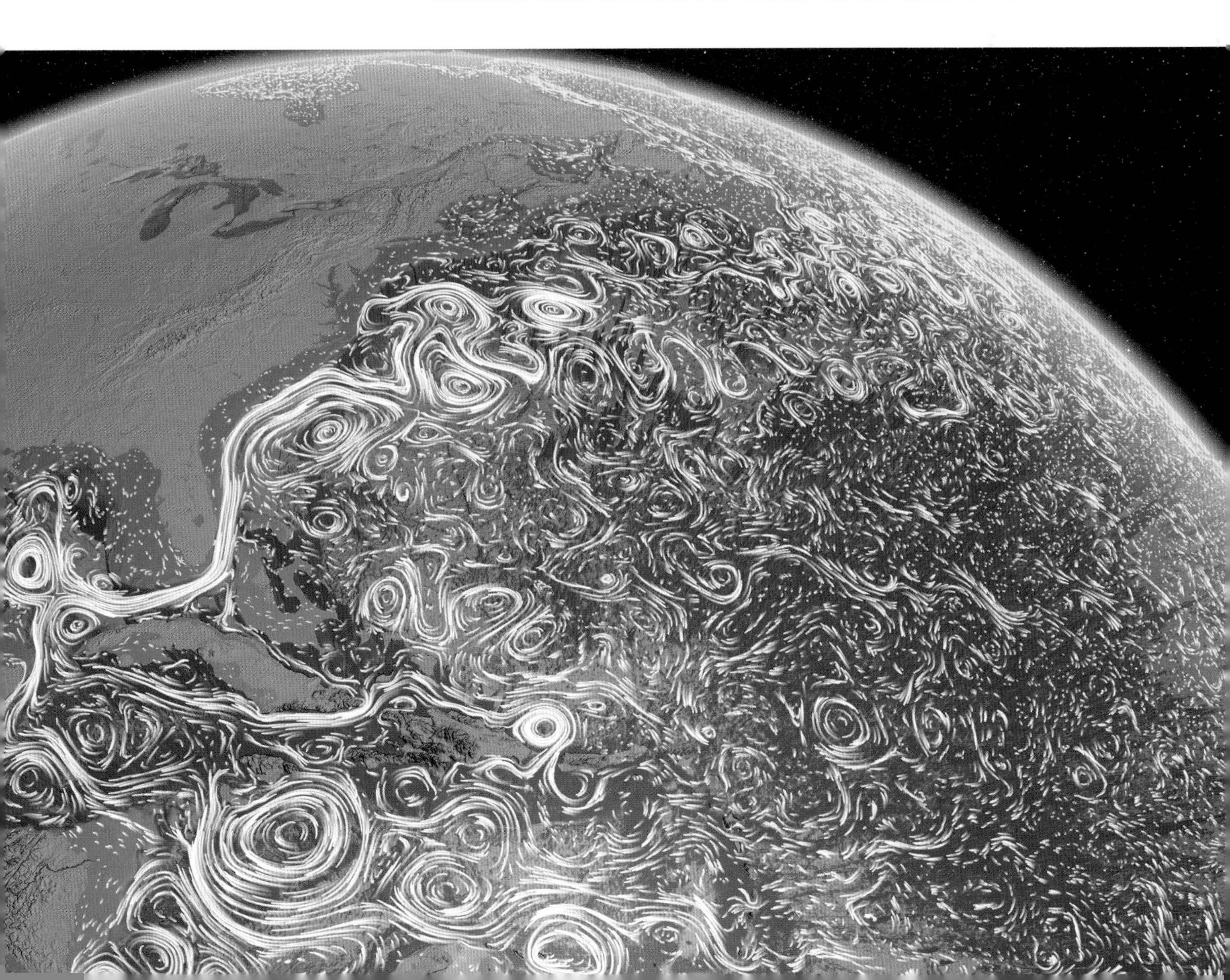

as cold as 60°F, very cold for seawater so close to the equator. The current causes upwelling of deeper, nutrient-rich and fish-rich waters near the coast of Peru—a great boon to the fishing industry there. Once every four years on average, the Humboldt Current weakens, as a result of changes in the wind that creates it. This is an "El Nino" event, and it devastates the Peruvian fishing industry, and also causes drought in Australia and parts of Asia and increased rainfall in East Africa. Another current that has a marked effect on climate is the Gulf Stream, which carries warm water out from the Gulf of Mexico and then continues northeastward across the Atlantic Ocean as the North Atlantic Drift. This current has a profound effect on temperatures in Northern Europe. For example, average winter temperatures in London, England, are around 20°F warmer than Calgary, Canada, which is at the same latitude but does not benefit from the influence of a warm ocean current.

Rising Up

Warmed seawater loses some of its energy when it evaporates, to become water vapor. When water evaporates, its molecules mix into the air one-by-one, darting around, tumbling, mixing, and colliding with the atoms and molecules already there. Water molecules constantly leave the surface of liquid water, and water molecules from the vapor constantly join or rejoin the liquid. At a higher temperature, the evaporation rate increases, since the average energy per water molecule is greater, and more molecules will have enough energy to escape the bulk liquid. The opposite is also true: at cooler temperatures, the evaporation rate slows, and water molecules already in the air are more sluggish on average, and are more likely to rejoin the bulk liquid if and when they hit. The pressure of the air is also important: at higher atmospheric pressure, the rate of condensation increases, since water molecules already in the air are more likely to be bumped back into the liquid by other water molecules or other air molecules. For each combination of temperature and pressure, then, there will be an equilibrium, at which the number of molecules leaving the liquid is equal to the number joining. At this equilibrium point, the vapor is said to be saturated. In a closed jar half-filled with water, the air will quickly reach saturation. But the air around you is rarely saturated: over a lake, for example, wind can carry water vapor away, bringing fresh, dry air into which more water molecules can evaporate; over a desert, there may not be enough water to saturate the air, whatever the temperature.

Humidity, simply the concentration of water in the air, can be expressed in units of pounds per cubic foot. Relative humidity, expressed as a percentage, is a measure of how the humidity of a sample of air compares with the humidity at saturation for that temperature and pressure. Relative humidity is 100 percent if the air is saturated. At 80°F and standard atmospheric pressure, a cubic foot of saturated air contains 0.0016 pounds of water (about one-fiftieth of an ounce). That means that a cubic mile of saturated air at sea level contains around 100,000 tons

Evaporation. In order to break free from the liquid, a particular molecule needs to have enough kinetic (movement) energy. Molecules of water possess a range of kinetic energies—the average determines (and is determined by) the water's temperature. Even at temperatures well below boiling point, some of the molecules will break free—and that number will increase with temperature. The escaped molecules mix with, and occasionally collide with, air molecules—predominantly nitrogen (blue) and oxygen (red). After colliding, some molecules are bumped back down to rejoin the liquid.

of water. One might suppose, therefore, that saturated air would weigh more than "dry" air. In fact the opposite is true. At any given temperature and pressure, a cubic mile of any gas contains the same number of particles overall. A cubic mile of dry air is composed almost completely of nitrogen and oxygen molecules, each of which weighs much more than a water molecule. So air containing water molecules is less dense than air without. It is no surprise, then, that warm, humid air is buoyed up by cool, dry air surrounding it. The dry air displaces the humid air, and itself warms up and becomes humid, and rises. Humid air can also be raised up if wind pushes it over a mountain, or if a mass of colder, denser air (a cold front) moving across the landscape uproots it. When air rises, it expands and cools—and as it cools, the water vapor will eventually reach saturation.

The temperature at which a given body of air is saturated is called the *dew point*. At the dew point, mist, fog or clouds will form in the air, and droplets of water—dew—will form on solid surfaces. An ice-cold drink on a summer day is below the dew point, and droplets form on the glass; droplets of dew will form on grass overnight if the air is humid enough and the temperature drops low enough. Water vapor can condense only onto a liquid or solid surface. In the air, solid surfaces are provided

by tiny particles called *cloud condensation nuclei*. In air without these particles, water vapor can be *supersaturated*, at up to 400 percent relative humidity, and still not condense. Typically, there are several hundred cloud condensation nuclei per cubic inch, mostly tiny specks of dust, soot particles from smoke, salt crystals thrown up by ocean spray, and even bacteria carried aloft by rising air. If the dew point temperature is below freezing (32°F), it is called the frost point, since water molecules will form ice, rather than liquid water, when they come out of the air. This process (the opposite of sublimation) is called *deposition*. Some clouds are made exclusively of water droplets, some only ice crystals, but most are a mixture of the two.

Water vapor holds the energy that evaporation takes from the ocean, and stores that energy, like a rechargeable atmospheric battery. If the sea surface temperature is consistently in the mid-70s Fahrenheit or above, as it often is in the tropics (near the equator), a tropical cyclone can form (called a hurricane when it forms in the Atlantic Ocean), and the dense clouds harbor the energy released when the water vapor condenses. These storms naturally rotate as air "falls" into the low pressure area at their core, and they continue gathering energy as they migrate away from the tropics, releasing the energy when they make landfall.

The results of condensation and deposition—water droplets or ice crystals built up one molecule at a time—can often be stunningly beautiful.

▲ **An astronaut** aboard the International Space Station captured this incredible view of Hurricane Florence making landfall on the Eastern Seaboard of the United States, in September 2018. *Source:* ESA/A.Gerst, CC BY-SA 3.0 IGO.

◀ **Under certain** conditions, light refracted by the ice crystals that make up a cirrus cloud creates a circumzenithal arc—often called an upside-down rainbow.

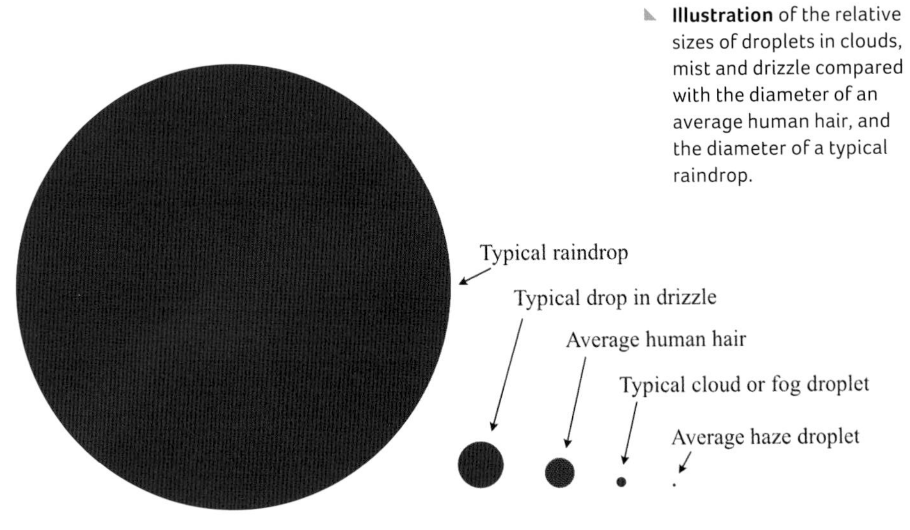

Illustration of the relative sizes of droplets in clouds, mist and drizzle compared with the diameter of an average human hair, and the diameter of a typical raindrop.

A single cumulonimbus, or thundercloud, can contain as much as 500 tons of water. Inside, rising air carries small ice crystals upward, causing them to crash into or rub against heavier ice crystals. As a result, the small crystals gain a positive electric charge, and the larger ones a negative charge, and the cloud gains an overall negative charge at its base and positive charge at its top. This separation of charge leads to the electrical discharge we know as lightning.

What Goes Up . . .

The ice crystals and water droplets that make up a cloud do not stay aloft forever. Sometimes they will evaporate again, causing clouds to disappear. Normally, however, they will fall back to Earth, as precipitation. There are many types of precipitation, and several different ways of classifying them, but the main ones are, of course, rain, snow, sleet (frozen or partly frozen rain), and hail.

Snow consists of snowflakes, each one an aggregation of several to many snow crystals—and a snow crystal is just a crystal of water ice that makes it to ground level. Most snow crystals actually begin as liquid water droplets. When the temperature inside the cloud drops well below freezing, a few droplets will freeze first, and grow by deposition of water molecules from the vapor around them. The droplets and the newly formed ice crystals constantly exchange water molecules with the air; evaporation is more rapid at the surface of a liquid droplet than at the ice crystal, so molecules tend to "leak" from droplets to crystals. The crystals grow, forming first hexagonal plates, and then afterward an array of different shapes, as more water molecules join, predominantly at corners and edges. A single snowflake is made of water molecules harvested from tens of thousands of water droplets. The shape of a particular ice crystal is determined by the variations in pressure, temperature and humidity during its growth. The delicate symmetry of snow crystals is explained by the fact that all its faces and edges experience the same environment as it grows. No two snow crystals experience precisely the same conditions,

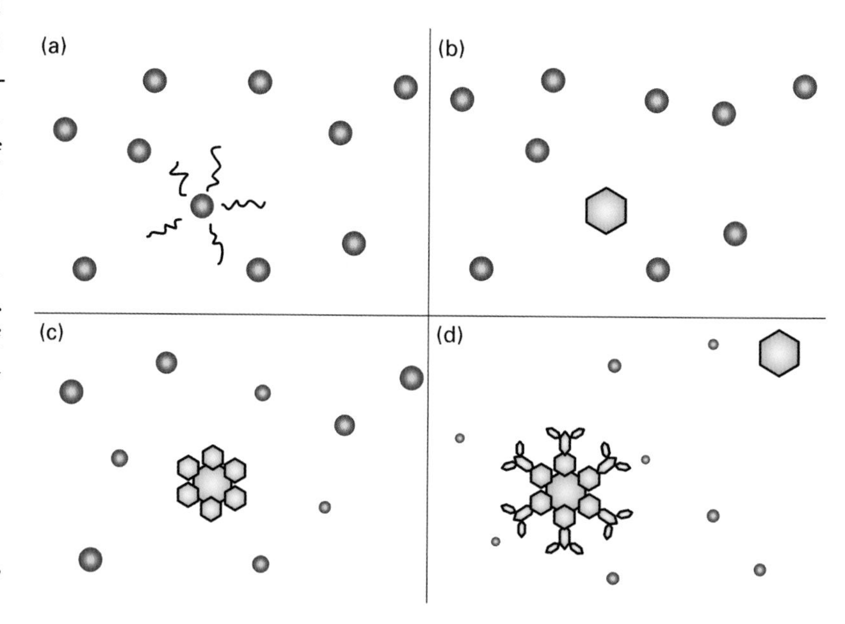

The beginnings of a snow crystal. When one water droplet freezes (a), it forms a small hexagonal plate (b). Liquid water droplets evaporate, and deposition of water molecules from the air is most likely at the corners, so the plate grows new branches there (c and d). The six-fold symmetry of the crystal reflects the way in which the molecules are arranged, which is explored in detail in chapter 4.

and each consists of trillions of water molecules—two facts that, together, explain why no two snowflakes can be identical. (It is worth noting that nearly identical "twinned" snow crystals can be grown under laboratory conditions.) American physicist Kenneth Libbrecht's book *Snow Crystals* provides intricate details of the processes involved in the formation of snow crystals, along with some stunning images.[10] Like many snow crystal enthusiasts, Libbrecht was inspired by the work of American farmer and amateur meteorologist Wilson Bentley, who pioneered snowflake photography, and Japanese physicist Ukichiro Nakaya, who pioneered the creation of artificial snow crystals and identified the various archetypes of snow crystals, and the conditions under which they form.

▲ **Wilson "Snowflake" Bentley** produced more than 5,000 photographs of snow crystals at his farm in Vermont. He caught snowflakes on a black velvet cloth, and photographed them quickly, before they could melt or sublime. *Source:* NOAA.

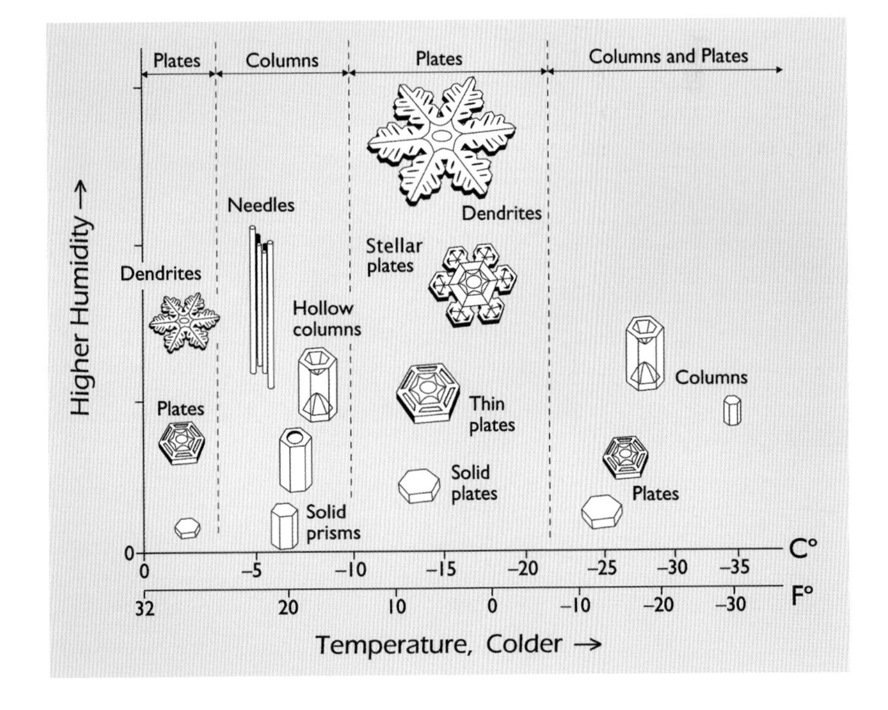

▶ **Japanese physicist** Ukichiro Nakaya was the first to create snow crystals in the laboratory, in order to study the processes that lead to their formation in clouds. As a result, he produced this diagram that shows how the production of different morphologies (kinds of shape) depends on temperature and relative humidity. This version was produced by Kenneth Libbrecht. *Source:* SnowCrystals.com.

▲ **This collection** of snowflakes photographed by Wilson Bentley hints at the tremendous diversity of forms inherent in snow crystals. *Source:* NOAA.

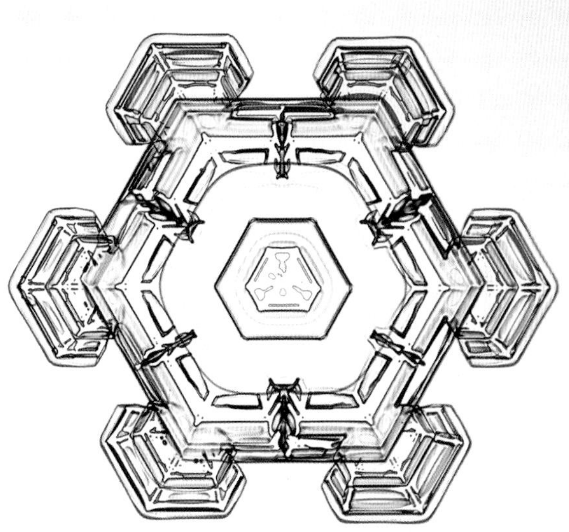

Two snow crystals, created and photographed by Kenneth Libbrecht. By varying the conditions under which the crystals grow, Libbrecht is able to produce endless different forms—and he takes photographs under colored light, adding an artistic touch. *Source:* SnowCrystals.com.

Electron microscopes provide intriguing views of snow crystals, showing details not visible in ordinary light micrographs. These images, of a stellar dendrite and a capped column, were captured at the Beltsville Agricultural Research Center, Maryland. The roughness on the capped column is rime, an accumulation of frozen water droplets. *Source:* Electron and Confocal Microscopy Laboratory, Agricultural Research Service, US Department of Agriculture.

Out in the Cold

In places where it is cold enough that falling snow does not melt—near the poles or at the top of tall mountains—it can persist for long periods of time. Snow falling on the tops of mountains builds up over time, compressing first to form *névé* and then, at the bottom, *firn*—a hard, granular form of ice. Eventually, the firn deforms under the weight of the snow and ice above it, and the whole mass oozes slowly downhill off the mountain: it becomes a glacier. Accumulation continues at the top, and at the bottom there is a *zone of ablation*, where ice melts or sublimes. As global temperatures increase, the rate of ablation has begun to dominate in many of these mountain glaciers, causing them to shrink back, or retreat. Glaciers that form on a mountain and are restricted to specific channels, like rivers of ice, are called alpine glaciers. Where a glacier covers a wider area, it is a called *continental glacier*. Medium-sized continental glaciers are *ice caps*; the very largest are *ice sheets*. Permanent ice sheets cover most of Greenland and Antarctica, where the residence time of a water molecule can run to millions of years. If the end, or terminus, of a glacier makes it all the way to the sea, chunks of ice "calve" off into the water, forming icebergs.

Seawater itself freezes at temperatures several degrees lower than pure water. When it does, it freezes out the salt, to become freshwater ice, increasing the salinity of the seawater around it. The cold, high salinity water is among the densest sea water on the planet, and it sinks, forming a crucial part of the thermohaline circulation—the global ocean currents below the surface, which are driven by changes in density caused by temperature (thermo-) and salinity (-haline).

Ice is always present in the Arctic Ocean, in the form of *ice pack*: small flat chunks and large islands of frozen fresh water. During winter in the northern hemisphere, around 15 percent of the area of the world's oceans (and therefore more than 10 percent of Earth's surface) is covered in sea ice. Just as alpine glaciers retreat as the climate warms, so do ice caps and the pack ice in the Arctic. Over the past few decades, Arctic pack ice has been shrinking year by year. This is a sign of our warming climate—but melting sea ice does not affect sea levels (just as the surface level of water with ice cubes floating in it does not change when the ice melts). However, when ice sheets and glaciers retreat, water that was previously on land enters the sea, and sea level does rise—a worrying trend for the next generation, since hundreds of millions of people around the world live close to current sea level.

Of course, over geological timescales, average global temperatures fall as well as rise, due to changing levels of atmospheric gases, changes in climate as a result of the movement of the

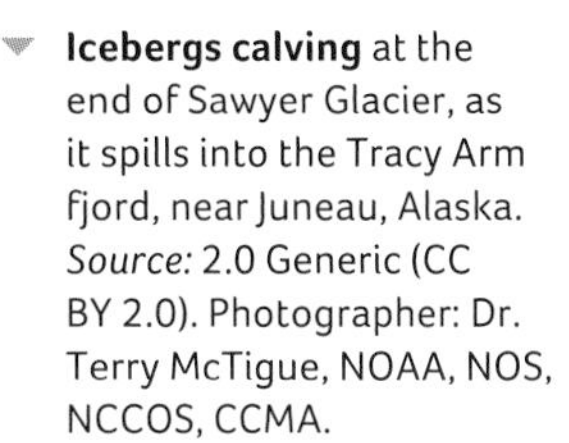

Icebergs calving at the end of Sawyer Glacier, as it spills into the Tracy Arm fjord, near Juneau, Alaska. *Source:* 2.0 Generic (CC BY 2.0). Photographer: Dr. Terry McTigue, NOAA, NOS, NCCOS, CCMA.

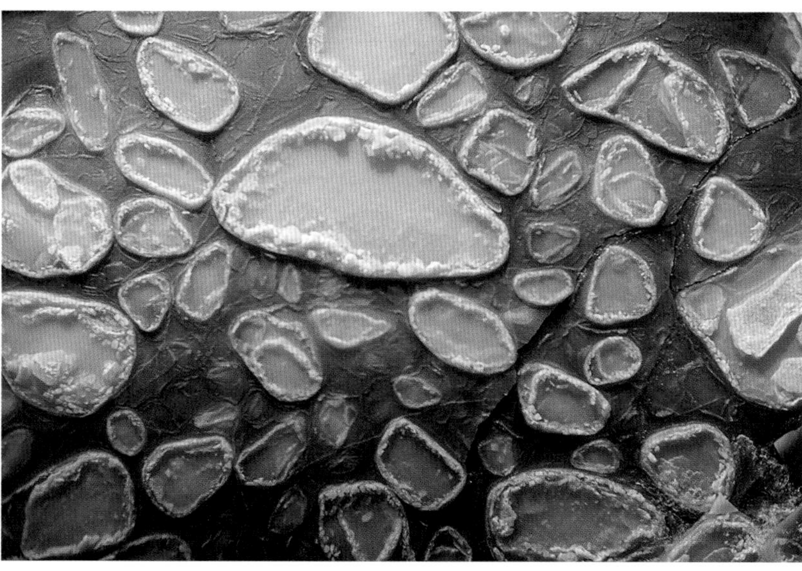

continents (plate tectonics), and long-term but regular fluctuations in Earth's orbit. When Earth's average temperature falls, the water cycle becomes unbalanced, and more snow falls, and stays, on the ice sheets and the extent of pack ice in the north grows. Technically, we are currently in a *warm interglacial period* of an ice age that began a couple million years ago and reached its peak about 20,000 years ago. During an ice age, ice sheets grow so much that sea levels can be tens or hundreds of feet lower than they are today. Around 30,000 years ago, the Bering Strait, which separates Russia and Alaska, became passable on foot—and when sea level was at its lowest, around 20,000 years ago, huge swathes of land were completely uncovered there. Several other "land bridges" were exposed in this way around the same time—including one that linked Britain to the rest of Europe. While the sea level was low at the height of the current ice age, the ice sheets extended much further than they do today, and were much thicker. As they retreated, they exposed glacial landscapes, on which the immensely heavy and slowly moving ice had gouged out valleys and lakes in Earth's crust.

NASA's Landsat 8 captured this stunning view of glaciers on the island of South Georgia. The two large glaciers that meet the water have both retreated by several miles since they were observed in 2000, a result of climate change. *Source:* NASA Earth Observatory images by Joshua Stevens, using Landsat data from the US Geological Survey.

Let It Rain

There is, of course, one form of precipitation that is more common than snow: rain. Around 121,000 cubic miles' worth of rain falls each year, mostly on the oceans. Some raindrops form when ice crystals melt as they drift downward, but most are produced when water droplets in a cloud coalesce to form larger droplets, which eventually become too heavy to remain suspended in the air. Small raindrops (just over 1/16 inch) are spherical in free flight, not teardrop-shaped as they are often depicted in cartoons or illustrations. Larger raindrops actually become flattened or even indented on the bottom as they fall, eventually resembling jellyfish or parachutes. The biggest raindrops on record were just short of 3/8 inch in diameter.[11]

When small flat pieces of sea ice bump together, their edges become raised, giving rise to this strange effect, called *pancake ice*. *Source:* Photographer: Michael Van Woert. Affiliation: NOAA NESDIS.

Like evaporation, rainfall varies over time and space. For example, one of the driest places on Earth is the Atacama Desert, in Chile, most of which receives less than 1/16 inch of rain each year. The Atacama Desert is in the *rain shadow* of two mountain ranges, and is also subjected to air that has been cooled (and dried) by the Humboldt Current. Some weather stations in the Atacama Desert have never recorded a single drop of rain. In complete contrast, Mawsynram, India, one of the wettest places on Earth, receives more than 450 inches of rain per year. The effect of the monsoon winds here is marked: the summer monsoon wind brings nearly 100 inches of rain in June but less than an inch in the whole of January. Mawsynram has nothing on a particularly rainy period that occurred around 230 million years ago. During the Carnian Pluvial Event, it rained more or less every day for two million years, causing extinctions for many plants and animals, and opportunities for others. In particular, it led to a dramatic diversification of the dinosaurs, allowing them to dominate our planet for millions of years.[12]

The village of Mawsynram, India. Officially the wettest place on Earth, with average annual rainfall of 467.4 inches. In 1985, Mawsynram received 26,000 millimeters (1,000 inches) of rainfall.

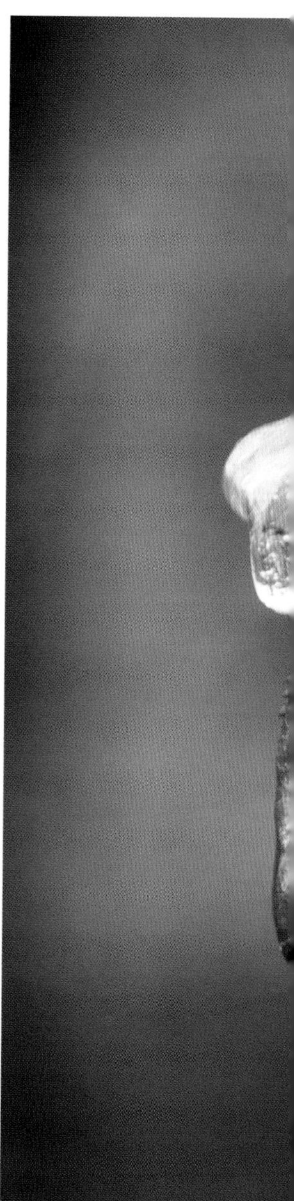

Freezing rain is made of raindrops that are supercooled: below freezing but still liquid. The drops freeze the moment they hit a surface—hence the name. *Source:* Image by susanne906 from Pixabay.

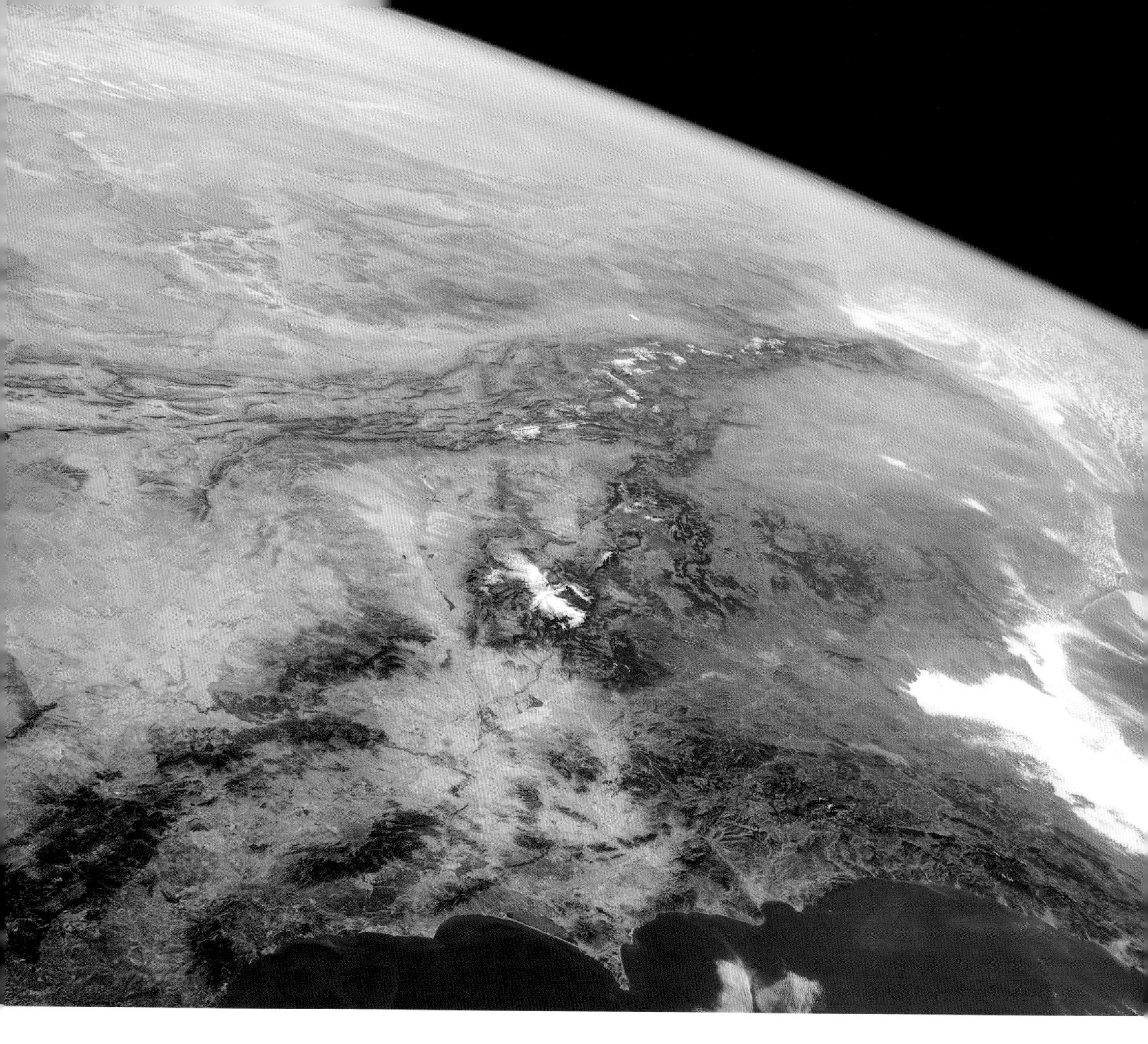

▲ **Rain shadow.** A prevailing westerly wind pushes humid air from the Atlantic Ocean over Morocco. The air has nowhere to go but up when it meets the Atlas Mountains. As it rises, it cools, forms clouds, and rain falls predominantly on the seaward side of the mountains as a result. This creates an arid region on the leeward side of the mountains—visible in this photograph taken by an astronaut onboard the International Space Station. *Source:* NASA's Earth Observatory.

Going Underground

Some of the rain that falls on the land (and some of the meltwater of glaciers and of the snow that has fallen on the land) evaporates straight back into the atmosphere. The rest continues being pulled downward by gravity, and ends up underground (as *infiltration*) or flowing off the land (as *runoff*). Some of the water that infiltrates into the soil will be absorbed into the roots of plants and lifted up into leaves, from where most of it will evaporate into the air—this process is called *transpiration* (see chapter 6). Any water not captured by roots will continue its downward journey, filtering through porous (permeable) soil and rocks until it hits a layer of water already there, called an *aquifer*. The water in an aquifer is prevented from sinking any lower by impermeable rocks below. The water in the aquifer is called *groundwater*, and the top surface of the aquifer is called the *water table*. In places where the ground's surface dips below the level of the water table, ponds and lakes will form. If the ground is sloping, the water may emerge, through a spring. Drilling down into an aquifer allows people to access water, via wells—the water lifted to the surface by pumps or buckets. In some places, groundwater flows into a space between two layers of impermeable rock, forming a *confined aquifer*. In that case, pressure builds up; drill a hole at an appropriate point in the topmost layer of rock, and you will make an *artesian well*, from which water will flood out at speed. If the geology is right, groundwater can emerge far from where it fell as rain: this is how oases form in deserts, for example.

Water is essential to human life, as it is to all living things. As hunter gatherers, our ancestors would have utilized water in the landscape—including some they found by digging underground. When people began to settle in small communities, they needed a reliable source from which they could draw continually. Evidence of the earliest purpose-built water wells dates back to around 10,000 years before the present.[13]

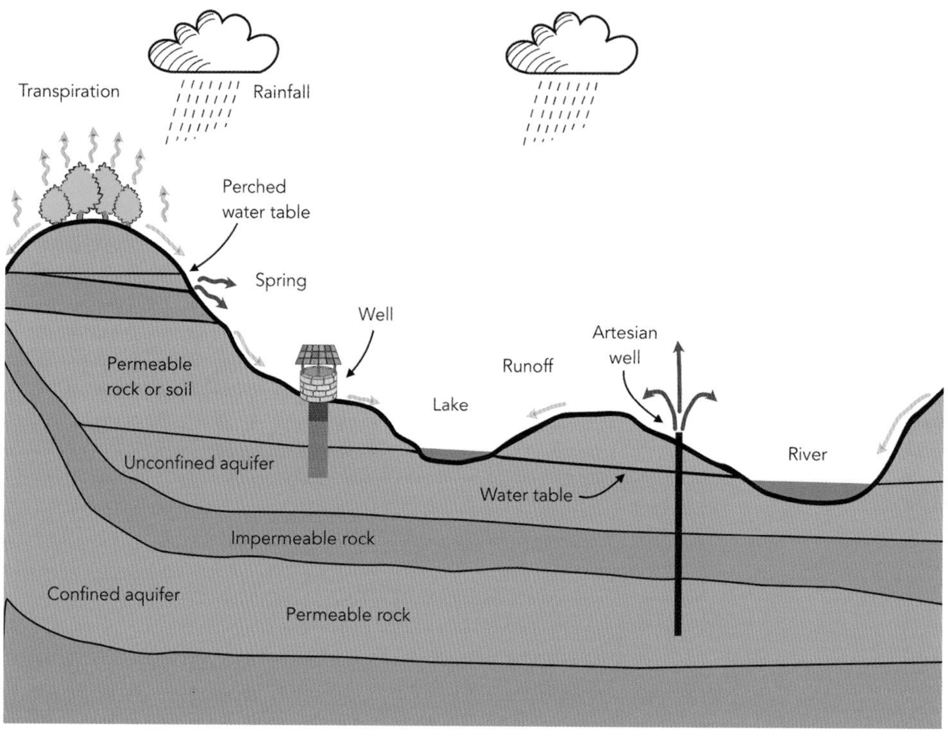
Transpiration
Rainfall
Perched
water table
Spring
Well
Artesian
well
Runoff
Permeable
rock or soil
Lake
River
Unconfined aquifer
Water table
Impermeable rock
Confined aquifer
Permeable rock

Let the Rivers Flow

When people began to develop larger, settled communities, they relied on rivers, since these watery veins carry much of the fresh water runoff and the infiltration from the land. Rivers meander their way to the sea, carving out their own routes through the landscape, growing larger, and carrying more water, the closer they are to the coast. One crucial advantage of a river compared with a well, for ancient civilizations at least, is that as well as supplying fresh water, rivers can also carry away the wastes that people create. And so, although the world's rivers contain only about 0.0002 percent of all the water at Earth's surface, they have played a crucial role in human history. The first civilization, founded around 5,500 years ago, was Mesopotamia, which began between the Tigris and Euphrates rivers in the Middle East (Mesopotamia means "between the rivers"). All the important ancient civilizations grew up around great rivers: the Egyptian civilization began in the Nile Valley around 5,000 years ago; the Indus Valley Civilization around the Indus Valley in India and Pakistan about 4,600 years ago; and the beginnings of the Ancient Chinese civilization in the basin of the Yellow River around 3,700 years ago (and the Yangtze River a little later). For their survival, people in each of these civilizations needed to learn how to manage water on a scale never seen before—how to obtain it in sufficient quantities from rivers and from wells, how to use it to irrigate their crops, how to drain it from land that was too sodden, how to store and transport it, and how to use it to wash away sewage and other wastes.

Illustration of water evaporation, transpiration, runoff, and infiltration.

Sumerian engineers in Mesopotamia pioneered the manipulation of levees[14]—steep banks at the edge of a river. Every spring, the Tigris and Euphrates carried with them huge amounts of silt, in meltwater from mountains far to the north. The torrents would shift and change the course of the river, breaking and remaking natural levees. The river's behavior had understandably catastrophic effects on the local agriculture. The engineers strengthened the natural levees, with reeds, mud bricks and bitumen, so that they would not shift, giving rise to a steady, reliable supply of

An artesian well in Egypt releases groundwater under pressure into an otherwise arid landscape. *Source:* CC 2.0: François Molle.

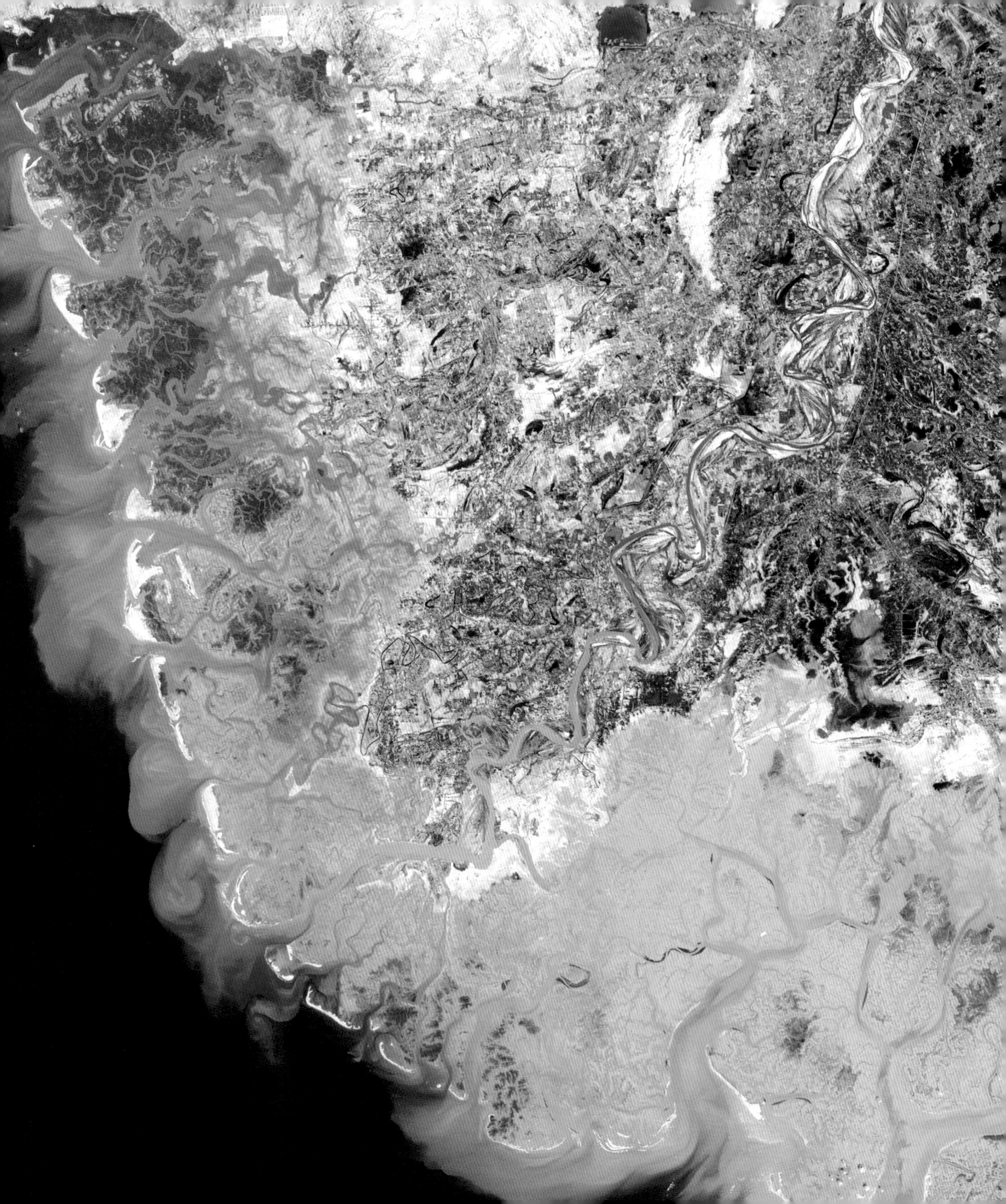

water. They also developed means by which they could lift water from the rivers and deposit it where crops grew, on the sloping sides of the levees and on the fields beyond. And they constructed canals that could divert the river water further afield. Canals also allowed people to transport goods around and between the various city-states of Mesopotamia.

Rainfall is highest between June and September in the basin of the Indus River, one of the cradles of civilization. Meltwater from glaciers in the Karakoram and Himalaya mountain ranges ensures that the river remains swollen for another couple months each year. NASA's Landsat 5 satellite captured this image—of the river and its extensive delta emptying into the Arabian Sea, in October 2009. *Source:* NASA Earth Observatory/Landsat 5.

In Ancient Egypt, the floodwaters of the annual inundation of the Nile Delta covered the agricultural lands with a thick layer of fertile mud, while the river itself sustained fish, birds and other animals that provided food. Egyptian engineers built dams and reservoirs to store water brought by the flooding, for use after the waters had receded.[15] The Indus Valley civilization is noted for its town and city planning, including sophisticated systems for water supply and drainage, built from standard-sized bricks in channels that were completely waterproof. The largest of the hundred or so settlements along the Indus River were Harappa and Mohenjo-Daro (both in what is now Pakistan) and Lothal (in what is now India). All three cities had hundreds of brick-lined wells, as well as water pipes, canals, reservoirs, and private and public baths. Most homes in Harappa and Mohenjo-Daro had private toilets—and all three cities had public toilets.

Artist's impression of Harappa, one of the cities of the Indus Valley Civilization, based on work done by the Harappa Archaeological Research Project. The city was planned on a grid system, and all houses were connected to a system of drains, which carried away waste and rain water. *Source:* Artist's conception of gateway at Mound E, Harappa. Drawing by Chris Sloan. Courtesy J. M. Kenoyer/Harappa.com.

Wastewater flowed through terracotta pipes into a network of brick-lined underground sewage pipes that carried the waste, and rainwater, away from the city.[16]

The cities of the Ancient Chinese civilizations also possessed canals and ditches to carry water into urban areas, and extensive earthenware and brick-lined sewers to carry wastewater out. Perhaps the most impressive water engineering in Ancient China was the Grand Canal—today a UNESCO World Heritage site. Work on this hydro-architectural marvel began about 2,500 years ago, and over the next thousand years or so, it was extended to its current length of more than 1,000 miles. It was one of many canals built across China, to join rivers and lakes to urban areas, mostly to facilitate trade. By the eleventh century, thousands of barges were carrying hundreds of tons of grain each year, northward from the south of the country. Over time, the canals became major conduits for tea, silk, porcelain and salt.

Troubles with Water

So dependent are we on water that its scarcity or mismanagement has led to the downfall of cities and even entire civilizations. For example, the people of the Indus Valley Civilization were forced to move away from the river basin on which they had depended, as a result of climate change. Weakening of the summer monsoon reduced the amount of water delivered to the Indus River, while at the same time winter rainfall on the foothills of the Himalayas proved attractive to farmers, who moved away from the city and into rural areas.[17] Similarly, severe droughts tested the ingenuity of the Maya people of Mesoamerica, in what is now Guatemala, Belize, and eastern Mexico. Mayan engineers built terraced drainage systems and extensive water storage facilities in an attempt to survive the droughts,[18] which often persisted for several years. But eventually, in the tenth century, drought made existing problems of overpopulation and warfare too much to bear, and many of the largest Mayan cities fell. Angkor, the capital of the ancient Khmer Empire, in what is now Cambodia, similarly had vast water storage and distribution systems—and it, too, succumbed to the perils of drought.

So vital is water that it has been at the center of conflicts since recorded history began—and probably earlier. The Pacific Institute maintains an online database[19] that summarizes violence related to "access to fresh water, attacks on water systems, the use of water as a weapon, terrorist incidents related to water, and more, going back nearly five thousand years."[20] The oldest conflict in which water was used as a weapon of sorts was part of an ongoing territorial battle between the city states of Umma and Lagash, in Mesopotamia, around 4,500 years ago. The king of fertile Lagash diverted rivers and canals in his region, depriving Umma of water. The Pacific Institute's database lists nearly a thousand situations, large and small, in which water supplies were destroyed, poisoned or fought over, or in which those water supplies suffered as a result of fighting. Conflicts continue today, often centered around water pollution, the construction of dams or the "mining" of groundwater. In various states of Australia, for example, farmers and other residents have organized pressure groups to battle against bottled water companies that have been buying up former agricultural land and using existing boreholes to drain groundwater—at the expense of nearby populations. There is public concern, too, about the perceived contamination of groundwater by fracking.[21]

Water constantly cycles between ocean, air and land, and we depend on it entirely for our survival. Natural variations in climate—a climate that is now changing rapidly as a result of human activities—can affect that cycle, bringing drought in some places and floods elsewhere. The Pacific Institute's database indicates that violent incidents over water have more than doubled in the past decade, compared with recent decades. Changing patterns of rainfall brought about by climate change are likely to exacerbate the situation in the decades to come. Peter Gleick, founding president of the Pacific Institute, commented for an article in the British newspaper *The Guardian*: "The evidence is clear that there's growing violence associated with fresh-water resources, both conflicts over access to water and especially attacks on civilian water systems." According to a 2018 study,[22] the main drivers of water conflicts are population density, water availability, the dynamics of the water flow in a river, power imbalance within a river basin's population—and climate. Rivers that are likely to experience "water wars" in the near future include the Ganges/Brahmaputra Delta (India), the Pearl (China), the Nile (Egypt), the Indus (India/Pakistan), the Colorado (United States), the Shatt al-Arab (a confluence of the Tigris and Euphrates in Iraq), and the Hari (Afghanistan). All of this while hundreds of millions of people still lack access to clean, safe water. The World Health Organization/UNICEF have jointly reported country, regional and global estimates of progress on drinking water, sanitation and hygiene (WASH) since 1990.[23] According to their 2019 figures:

- 1 in 3 people, or 2.2 billion people, around the world lack safe drinking water.
- Over half of the global population, or 4.2 billion people, lack safe sanitation.
- Almost half the world's schools do not have hand washing facilities with soap and water available to students.
- 2 out of 5 people, or 3 billion people, around the world lack basic hand washing facilities at home.
- 207 million people spent over 30 minutes per round trip to collect water from a source.
- Globally, at least 2 billion people use a drinking water source contaminated with feces.
- Some 297,000 children under five die annually from diarrheal diseases due to poor sanitation, poor hygiene or unsafe drinking water.

Leonardo da Vinci was certainly right in acknowledging that water is "sometimes health-giving, sometimes poisonous" and that "in time and with water, everything changes." Despite his lifelong fascination with water, Leonardo had no idea what it is made of. The next chapter examines how scientists came to realize that water is made of hydrogen and oxygen, and what we now know about the molecule that is H_2O.

CHAPTER 3

H_2O

In April 1783, Scottish engineer James Watt—of steam engine fame—sent his friend the English chemist Joseph Priestley a letter that he wanted read at the Royal Society, in London. In the letter, he wrote: “Are we not, then, authorized to conclude, that water is composed of dephlogisticated air and phlogiston . . . ?” Despite the archaic language, Watt was onto something. Until the 1780s, no one had considered water to be made of anything other than, well, water. What James Watt called “dephlogisticated air” was, in fact, oxygen, while “phlogiston” was hydrogen. Scientists soon realized that water is indeed a compound of these two elements—but it was not until 1860 that chemists determined the formula of water, H_2O. By the 1920s, scientists had discovered a special bond between water molecules—the hydrogen bond—that explains many of the behaviors that make water unique. By the twenty-first century, scientists could manipulate, and even split, individual water molecules.

Elementary Ideas

Great minds in all the ancient civilizations supposed that water is an element—a substance that is not composed of anything else, and so cannot be broken down into simpler components. It made sense: water does not visibly decompose into smaller, simpler parts (although we now know that it is constantly doing so—inside plants, for example). Ancient Indian, Chinese, Greek and Egyptian philosophers considered water one of the four (or, in some cases, five or six) elements from which everything is made. As an element, water was typically accompanied by air, earth and fire—though other combinations existed. Every unique substance was supposed to be composed of different combinations of the elements, and transformations of matter—what today we call chemical reactions—involved the transfer of elements from one substance to another. The existence of the classical elements could explain what happens when wood burns, for example. The

Burning wood, with the symbols most commonly used in Western alchemy to represent the "elements." Clockwise from top left: earth, air, water, fire.

wood must have contained fire, for that is released when it burns; it must have contained air, because you can see air rising with the smoke; it must have contained water, because the sap bubbles as the wood burns; and the ashy residue is mostly leftover earth.

These ideas—together with the suggestion that different elements had characteristic properties (water was cold and wet, for example; fire, hot and dry)—persisted and developed through the centuries. One practice that relied on theories of the elements was alchemy. People who practiced alchemy were attempting to understand and control the transformation of certain types of matter into others. It made sense that if all substances were made of the same four or five kinds of stuff, it should be possible to change one substance into any other. No surprise, then, that alchemical ideas permeated into metallurgy and medicine—areas in which alchemy's descendant, chemistry, still plays a central role.

Medieval woodcut of Persian alchemist Jabir ibn Hayyan, known in Europe as "Geber"—a key figure in Arabic alchemy in the eighth and ninth centuries. Geber recognized the importance of experimentation, and pioneered the use of many laboratory techniques still in use today, including many that involved water, such as distillation, evaporation and dissolving. *Source:* Wellcome Collection.

Alchemists in the Islamic Empire, from the eighth century onward, continued, added to and translated ancient traditions. They developed new methods and ideas, including adding in new "principles": sulfur, mercury and salt that would give substances properties such as combustibility, metallicity and hardness. Arabic alchemy heavily influenced practitioners in medieval Europe, and its ideas were entrenched in people's world views when the scientific method began to develop there in the seventeenth century. For example, despite his open-minded questioning of old ideas, even English scientist Isaac Newton dabbled in alchemy. Newton formulated a recipe for the "Philosopher's stone"—a substance that medieval alchemists believed could transform "base metals" such as lead into gold.

Through the many hundreds of years that alchemists used water in their procedures—dissolving, distilling, washing and mixing—never once did anyone think of it as anything other than elemental, incorruptible and unchanging.

A Burning Issue

The story of the discovery that water is, in fact, a compound begins with a theory about combustion that held sway among chemists for almost the whole of the eighteenth century. This new theory of matter was founded by German alchemist Georg Stahl, in 1703. According to Stahl, combustible materials contain a substance called *phlogiston*—and when these materials burn, they release some or all of their phlogiston. Stahl suggested that metals possess phlogiston, but metal *calces* (singular *calx*), found in ores, possess little or no phlogiston. Burn charcoal next to a metal calx—one way of smelting metals—and the phlogiston released restores the calx to the pure metal. Calces could also form over time—as rust, for example, as metals gradually gave up their phlogiston to the air. Many calces (without phlogiston)—rust, for example—appeared less dense than pure metal (with phlogiston). That made sense. However, it soon became clear that calces (today called metal oxides) are *more* heavy than their corresponding metals (because they contain oxygen). At the time, this posed a problem for some phlogistonists, especially in obvious cases, such as mercury: it is easy to convert the calx of mercury to pure mercury, and back again—and the calx always weighs more than the metal. Some ignored this challenge to Stahl's theory, while others suggested that phlogiston might actually have negative weight. Despite problems like this, phlogiston became a mainstay of chemical theories across most of Europe.

From the 1750s onward, scientists across Europe focused much of their attention on gases, which they called "airs." And phlogiston played a role in all the explanations of what they discovered—especially when those gases were combustible. In 1766, English aristocrat and amateur scientist Henry Cavendish discovered an extremely combustible gas: *inflammable air*, which today we call hydrogen. An invisible gas that burns was always going to be a challenge for the phlogistonists. Inevitably, perhaps, many scientists identified the gas as phlogiston itself—that is what James Watt was referring to in the letter quoted in the introduction to this chapter. Several scientists had made hydrogen before

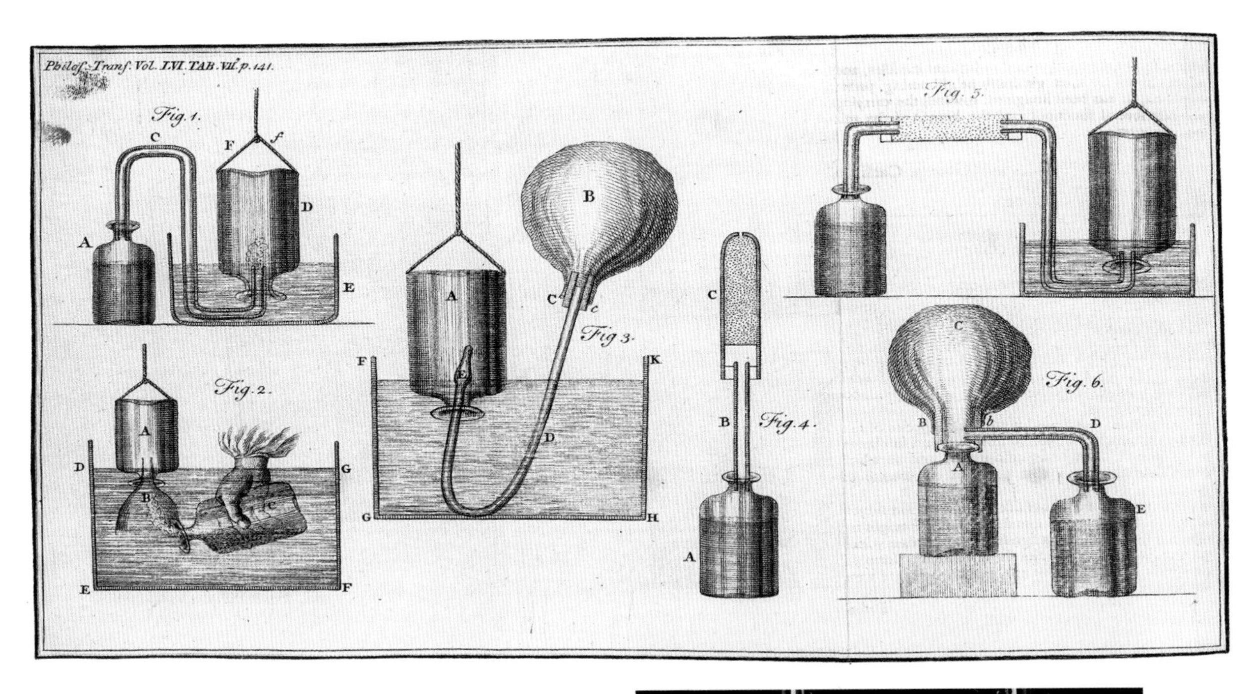

Henry Cavendish's typical experimental setup, taken from his 1766 paper. "In order to fill a bottle—with the air discharged from metals or alcaline [*sic*] substances by solution in acids, or from animal or vegetable substances by fermentation, I make use of the contrivance represented in Tab. VII."

A ribbon of zinc metal in dilute hydrochloric acid, which in Cavendish's time was known as "spirit of salt." The bubbles are pure hydrogen, just as Cavendish would have seen them.

Cavendish: it is produced whenever metals dissolve in acids (or, in some cases, even in water). But with his keen mind and thorough approach to experimentation, Cavendish was the first to explore its properties in detail. He described his findings in a paper presented to the Royal Society.[1] He wrote: "It has been observed by others, that, when a piece of lighted paper is applied to the mouth of a bottle containing a mixture of inflammable and common air, the air takes fire, and goes off with an explosion."

Each time Cavendish and his contemporaries caused inflammable air to explode, they were making water. But the water was escaping, as vapor, from the open ends of the jars and flasks they used to contain the gas, so they did not see it. The first inkling of the fact that the explosion of inflammable air produces water came when those explosions could be made to happen in closed containers. This was done inside a device called a spark eudiometer, invented in 1776 by Italian scientist Alessandro Volta (who would later invent the battery). In England, science lecturer John Warltire used Volta's device in his public demonstrations,

The most important scientists involved in the discovery of water as a compound. Left to right: Joseph Priestley, James Watt, Henry Cavendish and Antoine Lavoisier. *Source:* Wellcome Collection. Attribution 4.0 International (CC BY 4.0).

and noticed that each time he exploded inflammable air, the glass chamber was afterward coated with moisture. In a letter to his friend Joseph Priestley, he wrote: "Although the glass was clean and dry before, yet after firing the air it became dewey."[2] Warltire supposed that the moisture was deposited from the air, rather than being produced in a chemical reaction.[3]

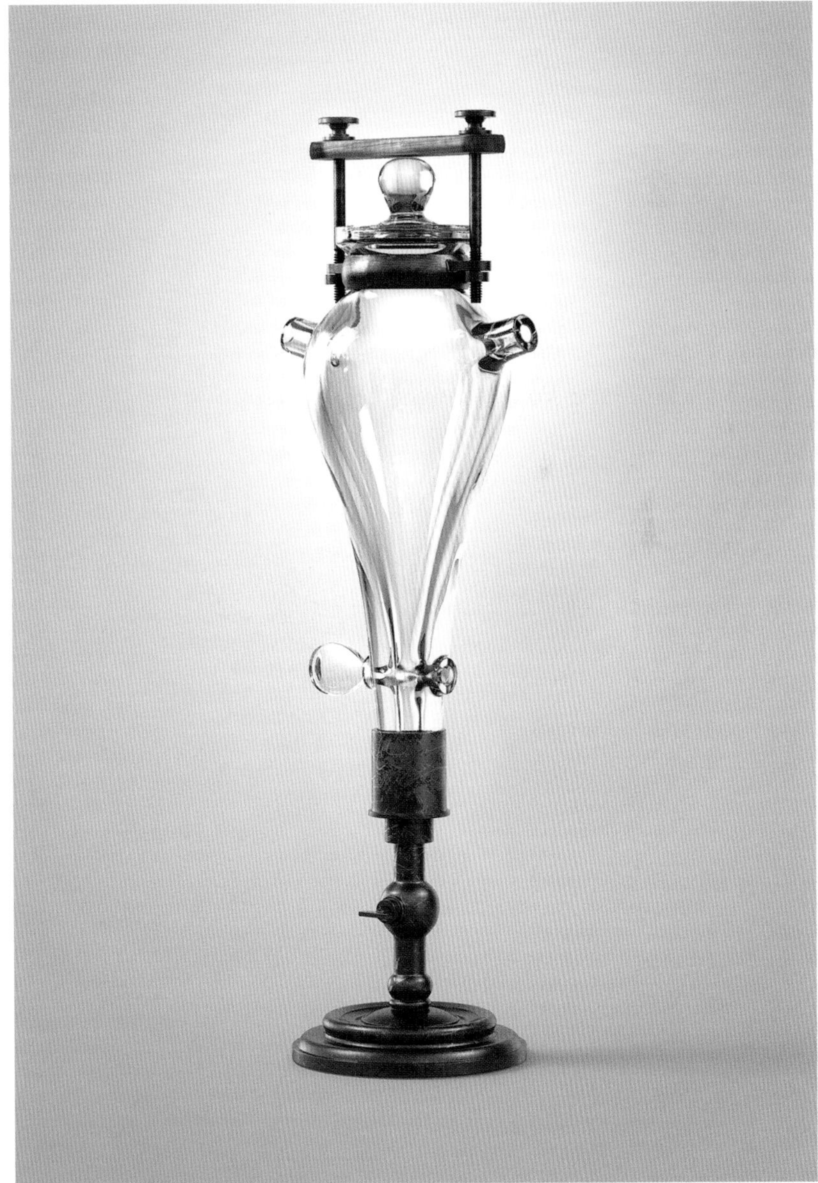

Artist's recreation of Cavendish's spark eudiometer. Wires poking in through corks plugging the openings in the sides of the glass container carried electric charge from a generator of static electricity (the battery would not be invented until 1799). A spark leaping from one wire to the other ignited a mixture of hydrogen and air inside the container. *Source:* Beautiful Chemistry.

Joseph Priestley was a key figure in *pneumatic chemistry* (the chemistry of gases)—it was to him that James Watt's letter was addressed. Priestley pioneered many of the techniques scientists used to investigate different "airs," and he is credited with discovering several new airs, including the one we now call oxygen. Again, others had already produced the gas, but Priestley was the first to publish about it. In a paper presented to the Royal

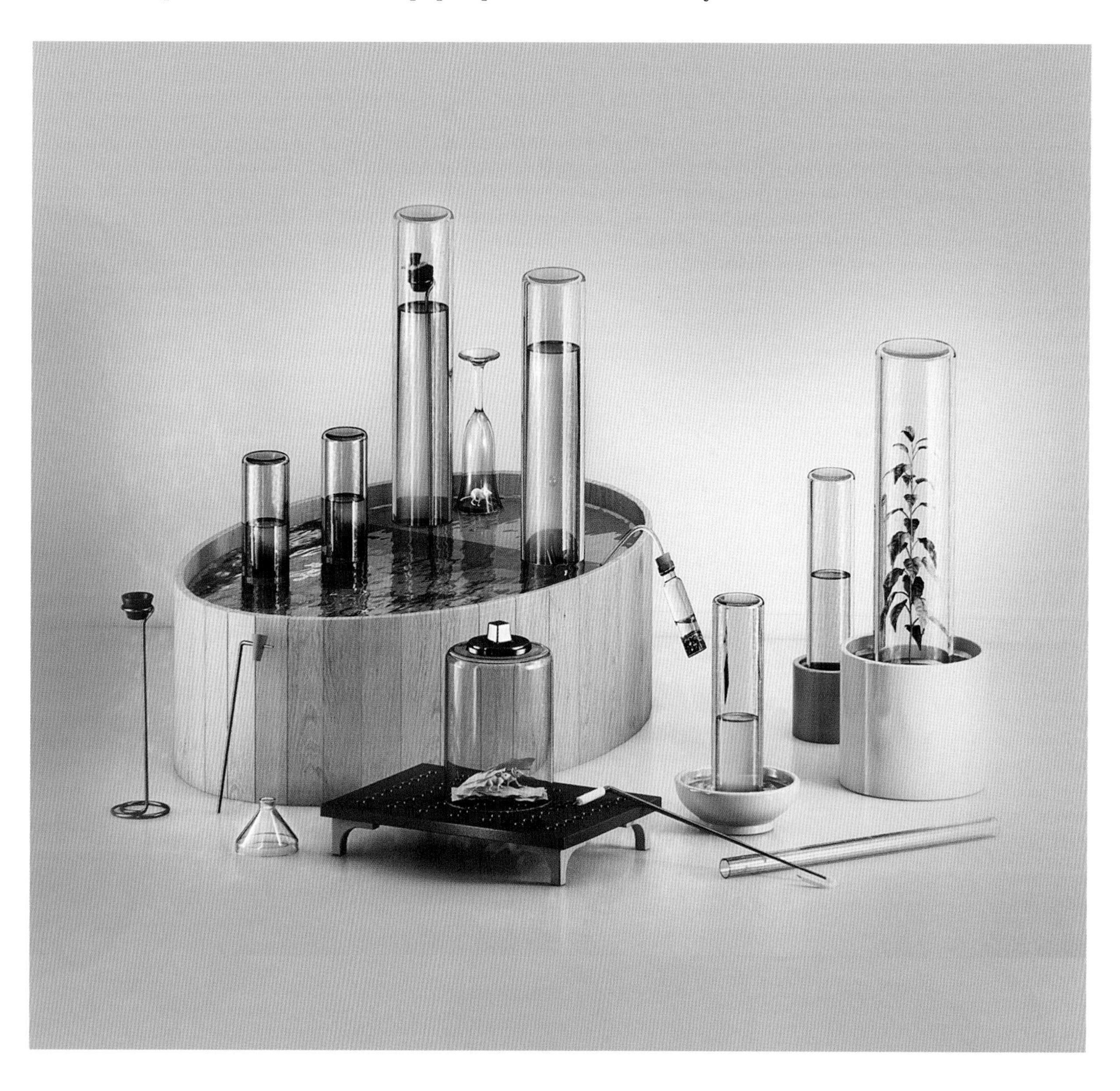

Society in 1775, he wrote: "the most remarkable of all the kinds of air that I have produced . . . is one that is five or six times better than common air, for the purpose of respiration, inflammation [burning], and, I believe, every other use of common atmospherical air. As I think I have sufficiently proved; that the fitness of air for respiration depends upon its capacity to receive the phlogiston exhaled from the lungs, this species may not improperly be called dephlogisticated." And that is why James Watt's letter referred to *dephlogisticated air* (oxygen).

Artist's recreation of Priestley's apparatus for studying "airs." Gases produced by plants and animals, and by chemical reactions, including burning, collected in the glass jars upturned in the water trough. *Source:* Beautiful Chemistry.

Revolutionary Ideas

Meanwhile in Paris, French chemist Antoine-Laurent de Lavoisier was busy laying the foundations of a revolution in chemistry (ironically, perhaps, he would later fall victim, at the guillotine, to the *political* revolution that was soon to begin in France). Lavoisier suggested that the calcination of a substance (combustion, rusting and respiration) involved that substance combining with a gas from the air. His fascination with combustion in particular began in 1772.[4] That year, he had burned sulfur and phosphorus in closed containers, and found that both elements gained weight. A few days later, he heated litharge (a compound we now call lead oxide) and found that it lost weight. He supposed that the same gas might be involved, combining with the sulfur and phosphorus and being released from the litharge. In 1774, Priestley had visited Lavoisier in Paris, and told him about his discovery of dephlogisticated air—and Lavoisier soon supposed that Priestley's gas might be the one he was looking for.

The products of burning sulfur and phosphorus formed acids when dissolved in water, and Lavoisier came up with the idea that the gas they were combining with was the cause. And so, in 1777, he suggested calling Priestley's gas *oxygène*, meaning "acid forming." When Lavoisier burned inflammable air (hydrogen) in oxygen, he was disappointed to find that the resulting liquid was not acidic, and so did not realize that the product of the reaction was pure water. Others did, and they soon surmised that water is somehow produced by the reaction of inflammable air (hydrogen) and dephlogisticated air (oxygen). Arguments over who first suggested this raged through the 1780s and 1790s, and again in the 1830s and 1840s—in two infamous episodes in the history of chemistry called the "water controversy." James Watt was probably the first to suggest the compound nature of water, in his letter of April 1783. The following year, Cavendish wrote that "we must suppose that water consists of inflammable air united to dephlogisticated air."[5] Cavendish and Lavoisier, among others, found that the water produced in the reaction weighed the same as the two gases that produced it. In 1783, Lavoisier suggested naming inflammable air *hydrogène*: "Thus water, besides the oxygen . . . contains another element . . . for which we must find an appropriate term. None we could think of seemed better adapted than the word *hydrogène*, which signifies the generative principle of water, from υδορ [hydros] aqua [water], and γεινομας [genes] gignor [come into being]."[6] In other words, hydrogen means "water-forming."

Lavoisier realized that if water is a compound of hydrogen and oxygen, he should be able to make the reaction go the other way—producing hydrogen and oxygen from the decomposition of water. He achieved it first by passing steam over red-hot iron filings; the high temperature gave water molecules enough energy to break apart, the resulting oxygen reacting with the iron to produce iron oxide. In a more controlled version of the same reaction, Lavoisier left iron filings in a closed vessel of water. Over a few

Drawing of Lavoisier's equipment for making hydrogen, taken from his 1789 textbook *Traité élémentaire de chimie*. Steam from the vessel on the right passes over hot iron inside a glass tube, and the water's oxygen reacts with the iron. The resulting hydrogen gas, plus any unreacted steam, passed through to the spiral condenser, so that the steam would become water. The hydrogen was collected in a jar off picture, through tube K.

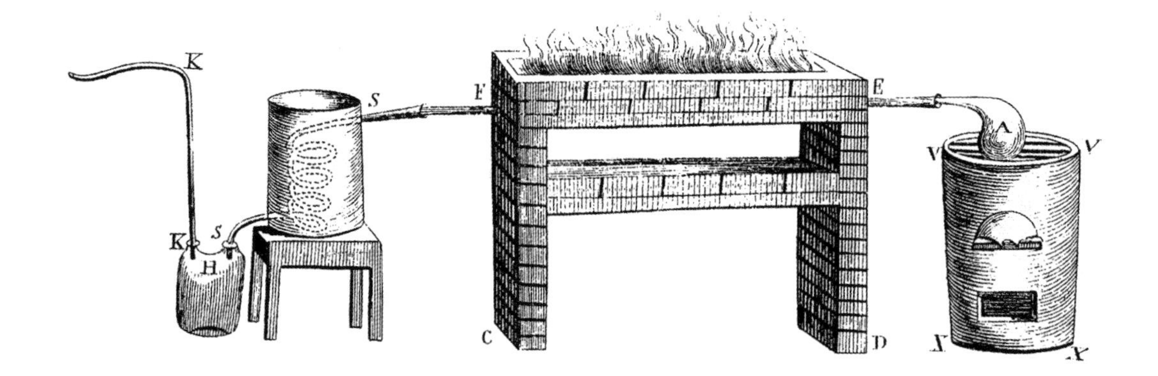

days, some of the water decomposed: the oxygen reacted with the iron to form iron oxide (a red powder), and a bubble of hydrogen gas appeared in the vessel. After measuring the volume of gas produced, Lavoisier concluded that water was composed of two-thirds hydrogen and one-third oxygen.

Current Affairs

Another preoccupation of scientists of the eighteenth century, in addition to gases, was electricity. It was not only employed to start the reaction between hydrogen and oxygen, in the spark eudiometer. It was also used to *decompose* water into its elements, adding further weight to water's compound nature. This was first done in 1789, using static electricity, by Dutch merchant Adriaan Paets van Troostwijk and his friend Johan Deiman. Inside a glass tube filled with water, a current of electricity produced by a powerful electrostatic generator flowed through the water, generating bubbles of gas. The gases collected at the top of the tube and were ignited by a spark through a separate set of electrodes. The gas disappeared, forming water so that the tube was once again full of water. The two researchers could repeat the procedure again and again.[7] In 1799, Alessandro Volta invented the electric battery—a pile of alternating copper and zinc discs interspersed with brine-soaked cloth or cardboard. The *Voltaic pile* produced a sustained, more or less constant, current of electricity. Within a few months, English surgeon Anthony Carlisle and English chemist William Nicholson built their own battery, and copied van Trootswijk and Deiman's experiment. In their initial experiments, the oxygen reacted with one of the electrodes supplying the current; using unreactive platinum electrodes instead, they were treated to a constant stream of bubbles at both electrodes.[8] They repeated the experiment with a more powerful battery and with water-filled tubes upturned in a bowl of water; now, they could collect the two gases separately.

Today, electrolysis is used on an industrial scale, to produce hydrogen—but it currently produces only around 2 percent of

Electric current passing through water, producing hydrogen and oxygen gases, in the ratio 2:1. To make this work well, it is necessary to dissolve a little salt or acid in the water, since pure water does not conduct electricity well.

the world's supply. That figure may well increase if and when the much anticipated *hydrogen economy* takes off, with cars powered by hydrogen replacing fossil fuel–burning cars that are commonplace today. Renewable electricity generated by solar panels or wind farms can be used to produce hydrogen from water, which can then be used as a fuel—the only waste product being water. Hydrogen cars are already on the market, and there is a small but growing number of hydrogen fueling stations, mostly in California and Japan. Despite the fact that hydrogen is very combustible, most hydrogen cars do not actually burn it, as regular automobile engines burn gas. Instead, in most hydrogen vehicles, hydrogen from a cylinder, and oxygen from the air, combine in a device called a fuel cell, which produces electricity from the energy released. Most hydrogen vehicles, then, are electric vehicles with their own onboard hydrogen-powered electric generators.

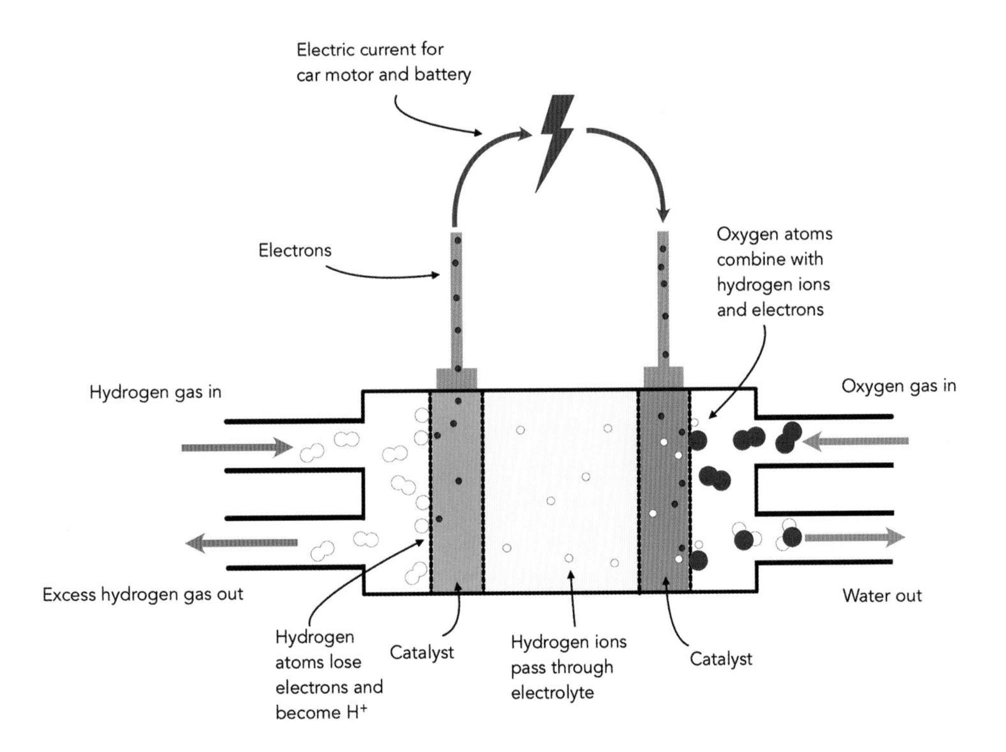

Schematic illustration of a fuel cell. A fuel cell has two sides: one fed with hydrogen gas under pressure, the other fed with oxygen (or air). In between are several layers, but the two most important are two catalyst layers separated by an electrolyte. The catalyst effectively lowers the *activation energy* needed to make the hydrogen and oxygen react. (Despite the reaction being energetically favorable, a mixture of hydrogen and oxygen will not spontaneously react.) In most systems, platinum is the catalyst. Hydrogen molecules adsorb (attach) to the platinum on one side of the electrolyte, and the hydrogen atoms split into protons (hydrogen ions, H^+) and electrons. The electrolyte allows only positive ions to permeate, and the protons pass across to the other side, where oxygen molecules from the air have adsorbed to platinum on the other catalyst. The electrons that have been separated from hydrogen atoms pass around a circuit—charging a battery or driving an electric motor. Oxygen atoms, protons and electrons combine on the other side of the fuel cell, forming water.

How Do You like Your Water? One H or Two?

Today, it is common knowledge that water is made of molecules, each consisting of one oxygen atom and two hydrogen atoms. But that was not obvious, even to chemists who had decomposed water into two volumes of hydrogen and one of oxygen. Although scientists in the late eighteenth and early nineteenth centuries mostly assumed that matter is made of tiny particles, there was no consistent theory of atoms and molecules—and no reason to suppose that water would be made of twice as many atoms of hydrogen as oxygen. The first modern atomic theory was proposed by English chemist John Dalton, in 1808. Dalton introduced the idea of atomic weight—he realized hydrogen was lightest, and thought oxygen was seven times as heavy (in fact, it is about sixteen times as heavy). He suggested that each molecule of water is made of one hydrogen atom and one oxygen: "there is the same number of in two measures of hydrogen as in one of oxygen."[9] So compelling was his atomic theory that his formula for water, HO, stuck.

In 1812, Italian scientist Amadeo Avogadro put forward a compelling case that equal volumes of any two gases at the same temperature and pressure consist of the same number of particles. This would have challenged Dalton's explanation of the two-to-one ratio of hydrogen and oxygen in water. But few people took notice of Avogadro, and Dalton's suggestion that water molecules were made of one H and one O persisted. The fact that hydrogen and oxygen gases are both diatomic (consisting of two atoms each, in this case H_2 and O_2) confused things a little—but Italian chemist Stanislao Canizarro worked it out in the late 1850s, and managed to convince other chemists that the formula for water should be H_2O, at an international conference in the German city of Karlsruhe in 1860. Finally, water was no longer an element, and every school kid could learn its formula: H_2O.

Charging Forward

The electrolysis of water heralded the beginning of a new scientific discipline: electrochemistry. Volta's battery enabled the discovery of elements such as sodium and potassium, and electroplating, in which metals at one electrode go into solution and are deposited on an object attached to the other electrode. These advances prompted English scientist Michael Faraday to define a new term, *ion*, for electrically charged particles that could pass through *aqueous* (water-based) solutions. In 1884, Swedish chemist Svante Arrhenius suggested that when solid crystals dissolve in water, they form pairs of oppositely charged ions. He also proposed, correctly, that water itself dissociates, into H^+ (hydrogen) and OH^- (hydroxide) ions. Arrhenius also suggested that an acid is a substance that, when dissolved, increased the number of hydrogen ions in the solution. So, for example, when hydrogen chloride (HCl) dissolves in water, to become hydrochloric acid, it dissociates into H^+ and Cl^- ions, increasing the concentration of hydrogen ions in the solution. In the 1890s, scientists developed a way to measure the concentration of hydrogen ions. In pure water—neither acidic or alkaline—one in 10 million (10^7) of the water's molecules is dissociated. This means that the concentration of hydrogen ions is $1/10^7$, or 10^{-7} (the concentration of OH^- ions is exactly the same, in the case of pure water, since all the H^+ and OH^- come from dissociation of water molecules).

In 1909, Swedish chemist Søren Sørensen developed a simple way of expressing the concentration of hydrogen ions in a solution, and therefore of how acid or alkaline it is: the *pH scale*. No one is quite sure what Sørensen intended the initial "p" to stand for, but chemists today tend to use the word "power" or "potential." Mathematically, it means "the negative of the base ten logarithm of the concentration"—which may not sound simple, but it is easily explained by considering the case of pure water: the hydrogen ion concentration in pure water is 10^{-7}, and the pH of pure water is 7. In a solution where the concentration of hydrogen ions is much higher, such as one in ten thousand (10^{-4}), the pH is lower—in this case, 4. The pH scale runs from 1 (very strong acids) to 14 (very strong alkalis).

We now know that hydrogen ions in pure water or aqueous solution do not float around as naked protons. Instead, each one attaches to a water molecule, forming a hydronium ion, H_3O^+. So, strictly speaking, pH is a measure of the concentration of hydronium ions, not hydrogen ions (and, strictly speaking, it is a measure not of the concentration of hydronium ions, but their *activity*—a closely related concept). In a further complication, hydronium ions are not stable entities—each one has a rather fleeting existence, as hydrogen ions jump from one water molecule to the next. So, while the total number of hydronium ions may remain constant in a particular solution, the extra positive charge—the proton, H^+—jumps extremely rapidly from one water molecule to the next.

This flitting of protons from one water molecule to the next is called the Grotthuss mechanism, after German chemist Theodor von Grotthuss, who—as long ago as 1805—suggested that water molecules dissociate under the influence of a battery, into negatively charged oxygen atoms and positively charged hydrogen atoms. Grotthuss reasoned that the battery's negative terminal (where the hydrogen gas is released) would pull a hydrogen atom from a nearby water molecule. The space left by the lost hydrogen atom would be taken up by a hydrogen atom from a neighboring water molecule. Hydrogen atoms would be passed along a line very quickly, from molecule to molecule, forming a conductive "wire" through the water, which connects from one side of the battery to the other in

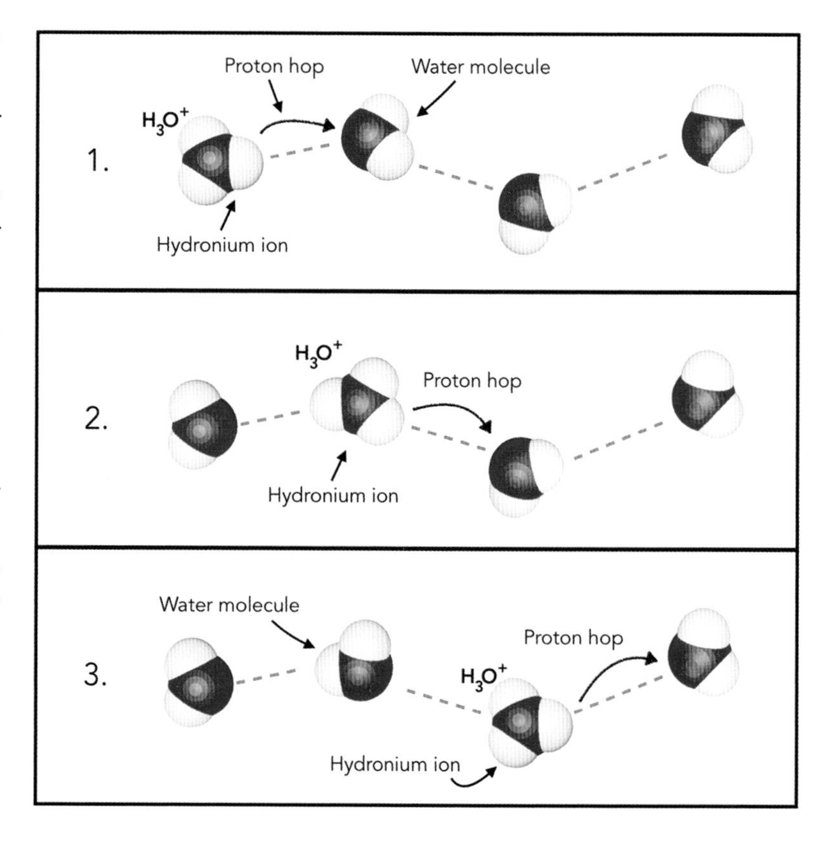

electrolysis—a little like the way a ball hitting one end of a Newton's cradle makes another ball move almost instantly. (Grotthuss also suggested that oxygen atoms do the same, in the opposite direction, which does not happen.)

The *proton hopping* in the Grotthuss mechanism is extremely rapid, and *water wires* really do form in the electrolysis of water. Metal ions in solution cannot take advantage of this mechanism: each one has to move physically through the solution—and it can do so only slowly, through the crowd of water molecules. This is a limiting factor in rechargeable batteries, which normally rely on the movement of (positively charged) lithium or nickel ions in the batteries' electrodes. With this in mind, in 2019, researchers at Oregon State University used a dye whose crystal structure holds water molecules—so that hydrogen ions, rather than metal ions, might be used to carry charge into the battery, at breakneck speed. The researchers say their work could "revolutionize electrochemical energy storage for high-power applications."[10]

Illustration of the Grotthuss mechanism. A hydrogen ion jumps from a hydronium ion H_3O^+ to a nearby water molecule. The hydronium ion becomes a water molecule, while the water molecule becomes a hydronium ion. The process repeats extremely quickly, with a different hydrogen atom each time—hundreds of trillions of times each second—effecting the transfer of positive charge through a "water wire." Dotted lines represent hydrogen bonds (see below).

The Shape of Water

In a water molecule, each hydrogen is joined to the oxygen atom by a *covalent bond*. Covalence involves sharing electrons. Each hydrogen atom has a single electron—but it would be more stable if it had two. In a hydrogen molecule (H_2), both atoms contribute their one electron to the covalent bond and benefit from the resulting stability. Similarly, an oxygen atom has six outer electrons, and reaches a stable configuration if it can have two more. An oxygen molecule (O_2) involves a double covalent bond, each atom sharing two of its electrons. In the covalent bonds that hold together a molecule of water, each hydrogen atom shares its single electron with the oxygen atom. And so, the reason why water is H_2O is that oxygen is well suited to sharing two electrons, and each hydrogen atom can supply one.

The covalent bond is strong—it takes a great deal of energy to break it. No wonder people believed water to be an element for so long. Those bonds can break as a result of heat energy: at 5,500°F, for example, a water molecule has about a 50 percent chance of splitting. In electrolysis, the energy of electricity breaks the bond, and in photosynthesis, it is the energy of sunlight—more on that in chapter 6. The dissociation of water molecules in liquid water, into and H^+ and OH^-, involves one of the covalent bonds breaking spontaneously; this has a very small chance of happening, which is why only one in ten million water molecules is dissociated in pure water at any one time.

The rise of quantum mechanics in the early twentieth century gave physicists a detailed knowledge of the structure of the water molecule. Quantum mechanics is the science that sets out to explain the behavior of extremely small objects and tiny amounts of energy. It models and predicts the behaviors of electrons, atomic nuclei and, indeed, complete atoms, ions and molecules—including how these particles and systems interact with each other, with light and with other forms of electromagnetic radiation. Quantum mechanics enabled physicists to work out the distances between the oxygen atom and its two attendant hydrogen atoms (the length of the covalent bond), as well as the overall shape of the molecule.

Water molecules are very, very small. When describing sizes and lengths at the atomic scale, scientists often use a unit called the *ångstrom* (Å). One ångstrom is one ten-millionth of a millimeter—which means there are 254 million ångstroms to the inch. The distance between the nucleus of the oxygen atom and the nucleus of each hydrogen atom is very close to 1 Å. As far as how large the diameter of an atom is, there is no single answer. This is because, in the quantum realm of very small objects, every entity behaves as both a particle and a wave. So, the electrons that surround the nucleus of each atom are "everywhere and nowhere." Rather than being solid particles in orbit around the nucleus, like planets around the Sun, they form diffuse three-dimensional *standing waves*. At each point of the wave, in three dimensions, there is a certain probability of the electron being found there at any moment. This probability tails off with increasing distance from the nucleus, so there is no definite boundary.

The most useful measure of the size of an atom or a molecule is the *van der Waals radius*, named after Dutch physicist Johannes van der Waals. When they are close (but not bonding), atoms and molecules repel each other, since the outermost electrons of both particles, being negatively charged, repel each other. As two atoms move closer, the repulsion grows, and when one atom's *van der Waals surface* touches the other's, the repulsion is very strong—almost as if the atom has a solid surface. This is why atoms bounce off each other (it is also why, despite the fact that every atom is more than 99 percent empty space, you don't fall through the floor). The van der Waals radius of an oxygen atom is 1.5 Å; for hydrogen, it is 1.2 Å. If you tried to push a hydrogen atom toward an oxygen atom, the two nuclei would never get closer than 2.7 Å (the sum of the two atoms' van der Waals radii). The covalent interaction between the hydrogen atoms and the oxygen atom enable a closer approach, and the water molecule approximates as a sphere with a van der

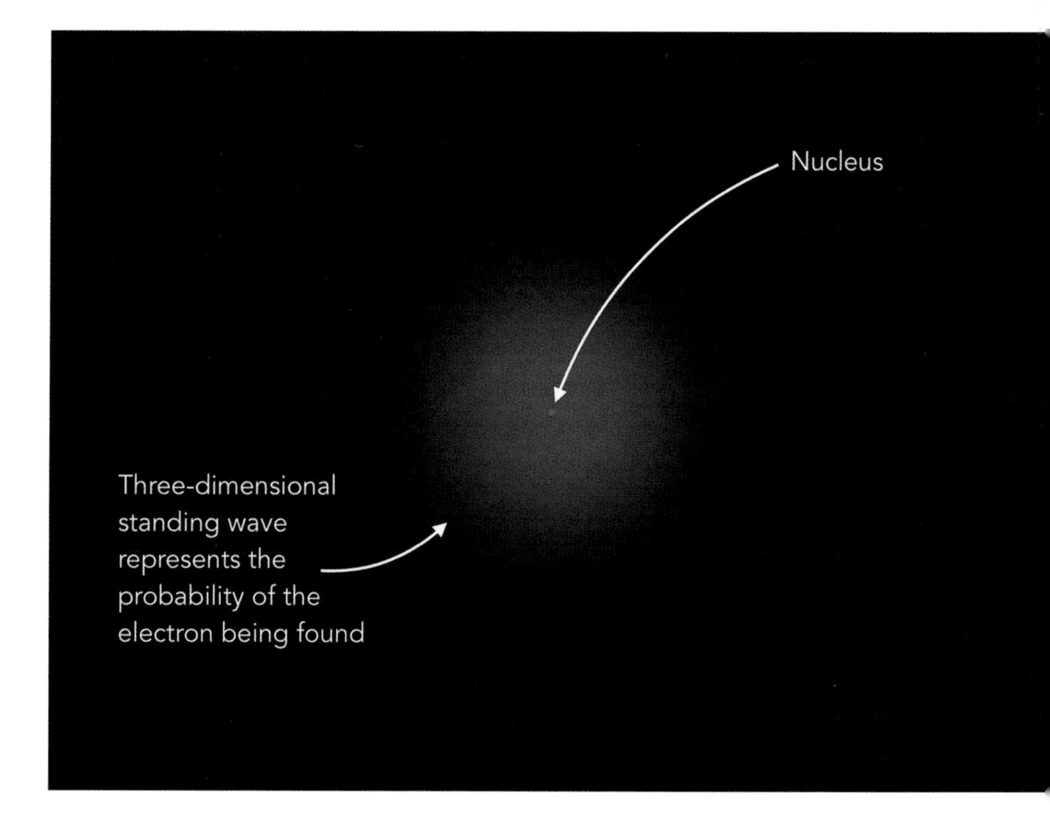

Illustration of a hydrogen atom as viewed by quantum mechanics, with no definite boundary. The atom's single electron is shown "spread out," as a three-dimensional wave that represents the probability of existing at each point in space. The nucleus is not to scale—it should be 1/100,000th the size of the atom.

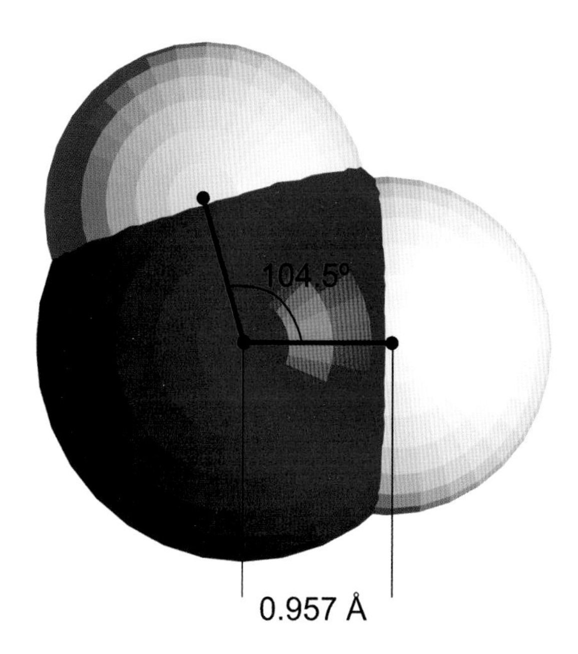

A representation of a water molecule with the three atoms shown as solid van der Waals spheres; locations of the nuclei are represented by black dots. Note how the distance between each hydrogen nucleus and the oxygen nucleus is much less than the sum of their van der Waals radii, thanks to the stability afforded by covalent bonding.

The near-tetrahedral shape of the water molecule, when including not only the covalent bonds but also the oxygen's lone pairs. The H-O-H angle is equal to 104.5°, slightly smaller than the angles in a regular tetrahedron, which are 109.5°.

Waals radius of 1.4 Å (the exact distance varies a little depending on circumstances, including whether there are other molecules nearby).

The van der Waals "surface" of the water molecule is not quite a sphere. This is because the covalent bonds do not lie in a straight line. There is an angle between them of 104.5° (the exact angle varies a little, again depending on whether there are other molecules nearby). The water molecule is bent, like a boomerang. Picturing the van der Waals surfaces of the bonded atoms as the molecule's overall form, the water molecule is often compared to Walt Disney's character Mickey Mouse (the oxygen atom is the head, the two hydrogen atoms the ears). The reason the molecule is bent is that the oxygen atom has four outer electrons that are not involved in bonding. These electrons are housed in two regions called *lone pairs*. The negative charge of the electrons in the lone pairs repels the negative charges of the covalent bonds, which are also rich in negative charge. And the two lone pairs also push each other apart. The result is that the water molecule approximates to a regular tetrahedron (a shape with

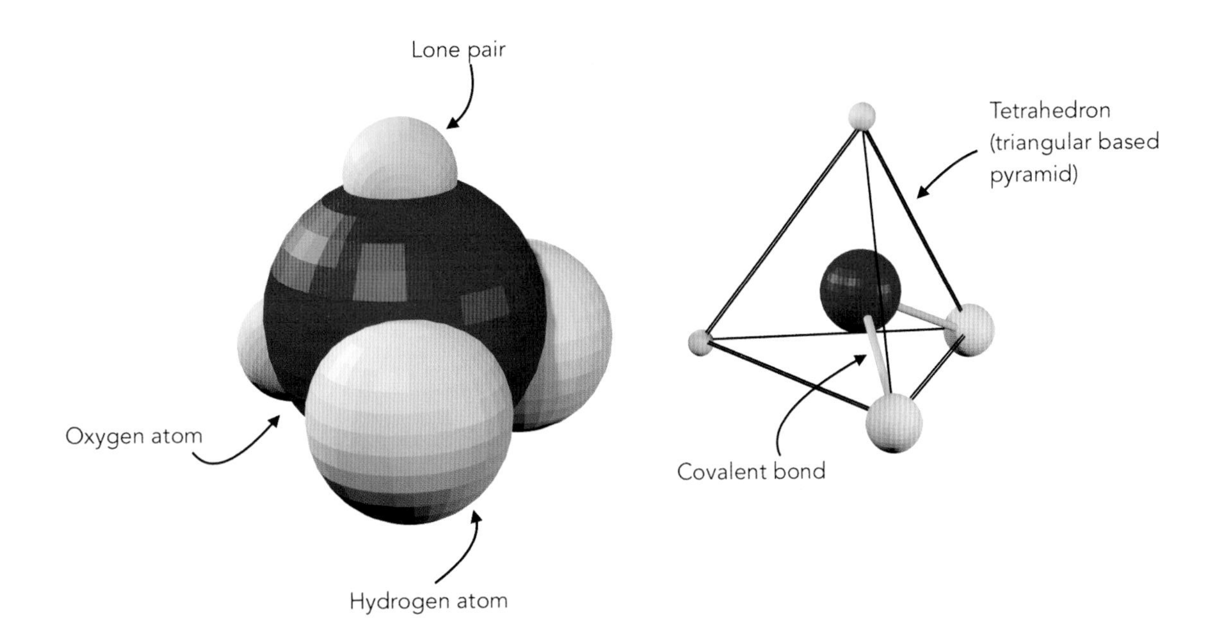

four identical faces, a triangular-based pyramid), with the two hydrogen atoms at two of the vertices (corners) and the two lone pairs at the others.

A Special Bond

Water molecules carry no overall electric charge, since the total number of (positively charged) protons in the atoms' nuclei is equal to the total number of (negatively charged) electrons: eight in the oxygen atom and one each for the hydrogens. However, the electric charge is concentrated in certain parts of the molecule. There are concentrations of negative charge in the oxygen atom's lone pairs, each of which contains two electrons in confined spaces. Meanwhile, the negative charge is more spread out around the hydrogen atoms: the electrons involved in the covalent bond are just as likely to be close to the oxygen atom as they are to the hydrogen, and they leave the hydrogen nucleus a little exposed. As a result, the oxygen atom is overall slightly more negatively charged than the hydrogen atoms. The water molecule has *partial charges*: partial negative (δ^-) at the oxygen and a partial positive (δ^+) at each of the hydrogens. This gives water molecules a *dipole moment*; water molecules are *polar molecules*—and in an electric field, they will orient themselves accordingly. This fact can be observed by holding an electrically charged object near a stream of water from a faucet; the water is attracted to the charged object, and the stream diverted. The polarity of water molecules is also responsible for a fascinating phenomenon, called the water thread experiment, in which water forms a flowing, unsupported bridge between two water-filled beakers held up to an inch apart. The bridge forms when high-voltage electrodes of opposite signs are placed into the two beakers.

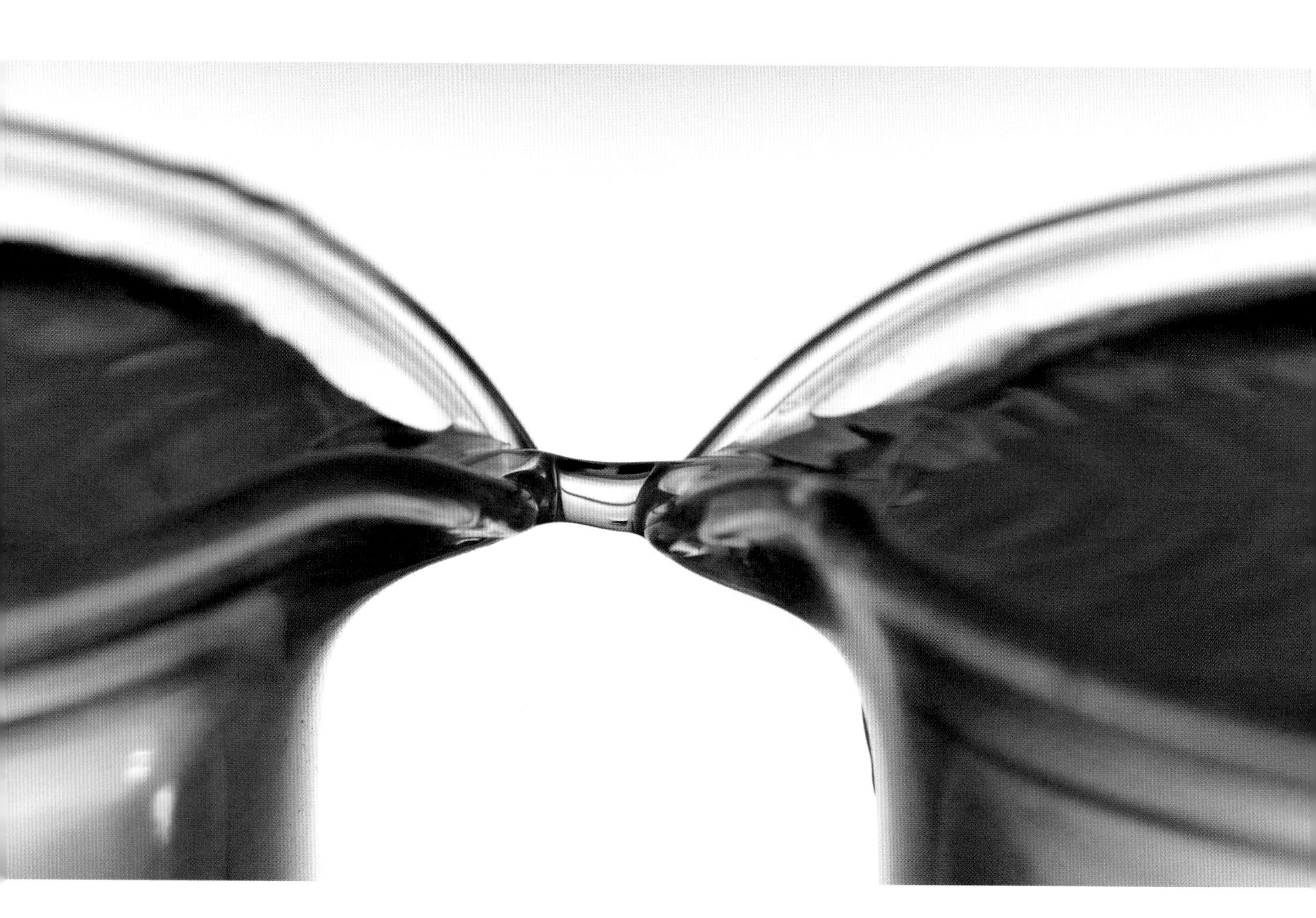

The unequal charge distribution in a water molecule gives rise to an electrostatic interaction between neighboring water molecules (and between water and other atoms, ions and molecules, too). This interaction is called a *hydrogen bond*. As we shall see, hydrogen bonding is behind several of the unusual properties of water, including water's high boiling and melting points, and the fact that ice floats on water (see chapter 4). It also explains surface tension, capillary action and water's ability to dissolve so many substances (see chapter 5).

The partially charged regions of positive and negative charge of neighboring water molecules can attract and repel each other. Water molecules rotate so that the partial negative part of one water molecule (one of the oxygen atom's lone pairs) lines up with the partial positive of another (one of the hydrogen atoms), and is attracted toward it. A hydrogen bond is not as strong as the covalent bond, but it is strong enough to bring adjacent water molecules closer (2.8 Å) than twice the van der Waals radii of an oxygen atom (3.0 Å)—closer than two isolated oxygen atoms could approach each other. Since a hydrogen bond forms between a hydrogen atom and one of the lone pairs of the oxygen atom, any water molecule can bond with four other water molecules: two at its hydrogen atoms (where it is a *donor* of the hydrogen bond) and two more at the oxygen lone pairs (where it is an *acceptor* of the hydrogen bond). And since the lone pairs and hydrogen atoms of a water molecule form a near-tetrahedral shape, so too do the hydrogen bonds that extend out from those points. The tetrahedral arrangement of nearby molecules is particularly important in ice—where the water molecules are in fixed positions, held together by hydrogen bonds—but it is also important in liquid water, where molecules form hydrogen-bonded groups called *clusters* (see chapter 4).

The water thread experiment. When a strong voltage (several kilovolts) is applied between bodies of water in separate beakers, the water flows from one to the other across a self-supporting bridge. No satisfactory explanation exists to account for this surprising phenomenon, although it is almost certain that the polar nature of water's molecules is somehow the cause. *Source:* Image by Gmaxwell, published under a Creative Commons GNU Free Documentation License Version 1.2.

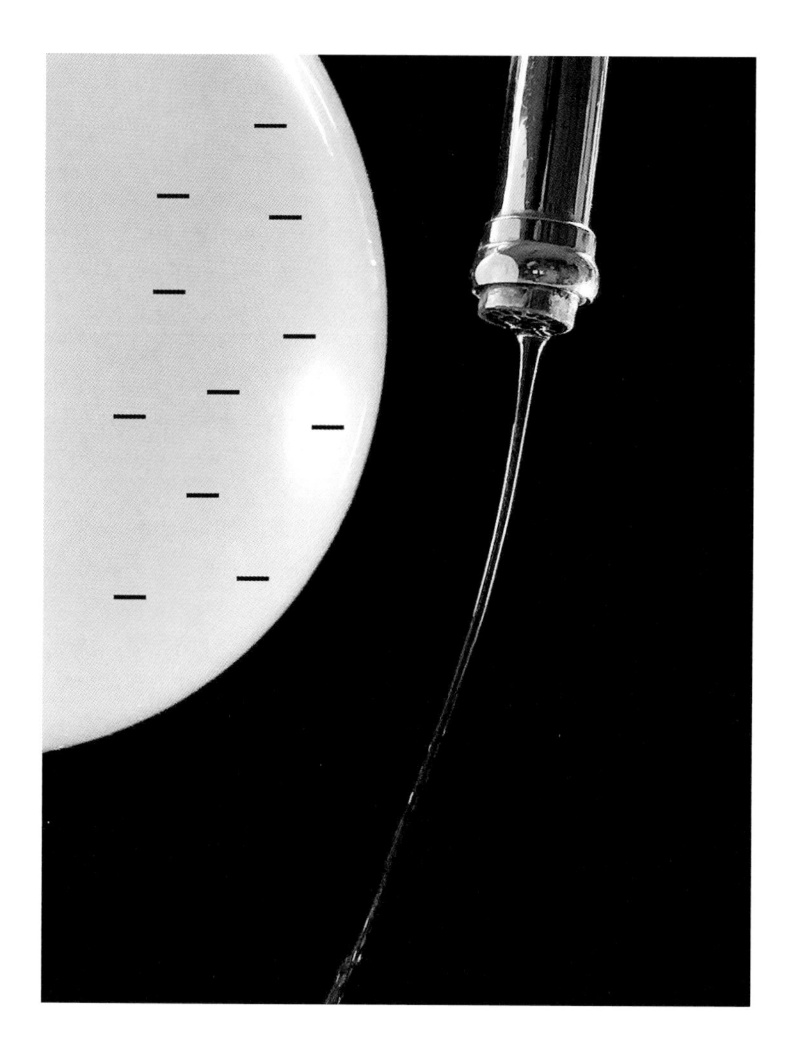

A balloon carrying a negative static electric charge attracts a stream of water from a faucet. Inside the stream, the hydrogen atoms of each water molecule, which carry a partial negative charge, are attracted to the balloon.

The partial positive and negative charges of two neighboring water molecules are attracted to each other, forming a hydrogen bond. The bond brings the two molecules closer within each other's van der Waals surfaces.

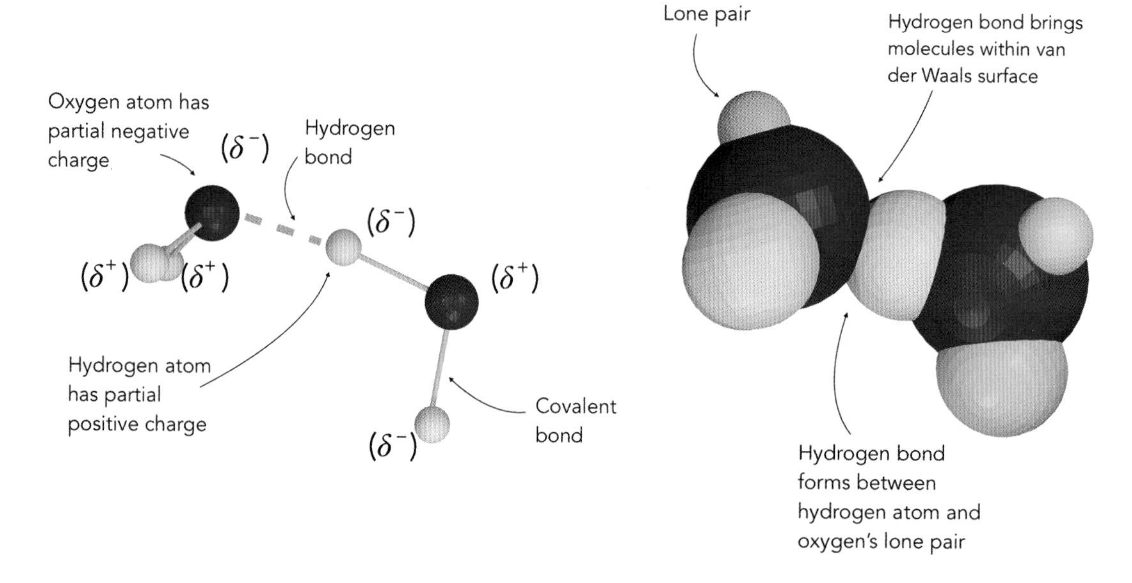

Bend, Stretch and Twist

Covalent bonds are not rigid: they can be thought of as springs or pieces of elastic. When electromagnetic radiation hits a water molecule (as infrared or ultraviolet light, for example), it can set the bonds stretching and relaxing. Incoming electromagnetic radiation is a varying electromagnetic field, so it interacts with the partially charged parts of the molecule. Electromagnetic radiation interacts strongly with all polar molecules, not just water. Radiation can also make the bonds flex up and down, in a scissoring action, so that the 104.5° angle repeatedly decreases and increases. In both these processes, the molecule absorbs the energy of the radiation. Certain frequencies of electromagnetic radiation will make this happen much more than others—just as pushing someone on a swing at the "natural frequency" of the swing will make the swing fly higher and higher.

There are three different "modes" of vibration (bond stretching and flexing), and three more of a reciprocating back-and-forth part-rotation, called *libration* (see page 100). Each of the six modes can be activated by various frequencies. Since the energy of the radiation is absorbed when this happens, water (and any other polar molecule) has a characteristic *absorption spectrum*. Knowledge of the absorption spectrum of water enables astrophysicists to detect water in space, where those molecules absorb radiation from nearby stars. (And since the energies absorbed are different if one or both of the hydrogen atoms is deuterium—hydrogen-2—astronomers are able to differentiate between hydrogen and deuterium far out in space. See chapter 1.)

In liquid and solid water, the molecules cannot rotate, but they do still stretch and flex, and some of the radiation they absorb is visible light—mostly at the red end of the visible spectrum. The loss of some light from the red end of the spectrum accounts for the pale turquoise blue color of liquid water and ice—although the weak absorption means that the color is visible only in large areas of water, such as the ocean or expanses of glacial ice.

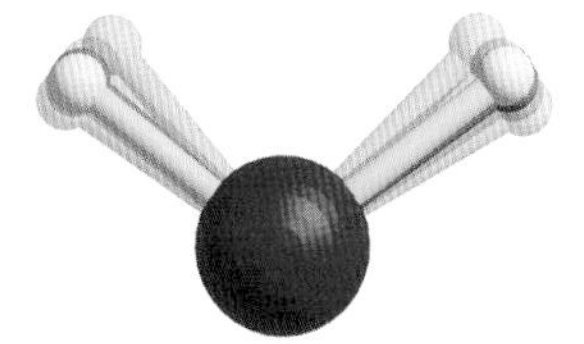

v_1
Bonds bend inwards and outwards

v_2
Bonds stretch and relax symmetrically

v_3
Bonds stretch and relax asymmetrically

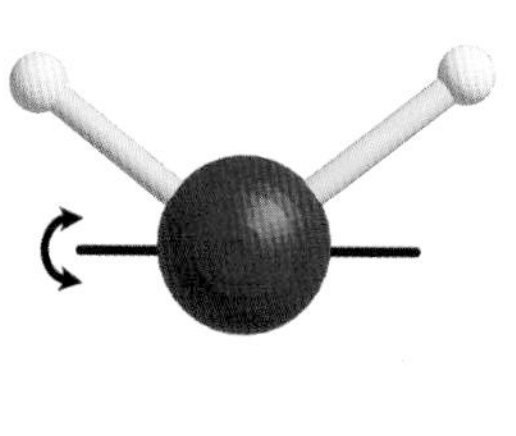

x
Molecule librates back and forth around x-axis

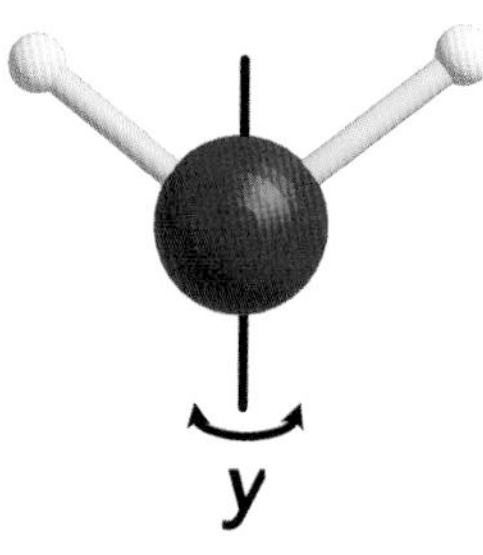

y
Molecule librates back and forth around y-axis

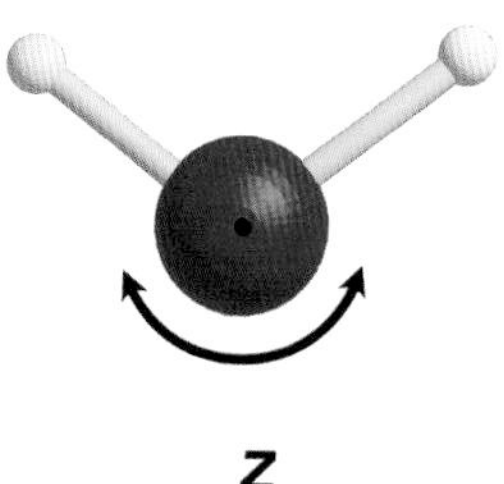

z
Molecule librates back and forth around z-axis

Illustration of the vibrational and librational (incomplete backward-and-forward rotation) modes water molecules can adopt when they absorb energy from electromagnetic radiation.

Liquid water really is blue—and so is ice. Water molecules absorb some frequencies of visible light, predominantly at the red end of the spectrum.
As a result, the color of water becomes bluer the deeper it is.

A Closer View

Atoms and molecules are far too small to see directly with ordinary microscopes. Fortunately, various methods are available for producing faithful images of these tiny objects. In particular, *scanning probe microscopy* involves an extremely sharp tip (only a few atoms across at the sharp end) passing across a surface, following the contours of the electron standing waves of atoms and molecules. There are two main kinds of scanning probe microscopes: the *scanning tunneling microscope* (STM, invented in 1981) and the *atomic force microscope* (AFM, invented in 1985). Both kinds work best with materials that conduct electricity and whose atoms are in regular arrangements; both have produced stunning images of metal atoms at a surface. It has proved more challenging to obtain images of water molecules—not least because the electrified tip of the devices can disrupt the water molecules—but in recent years, some researchers have managed it.

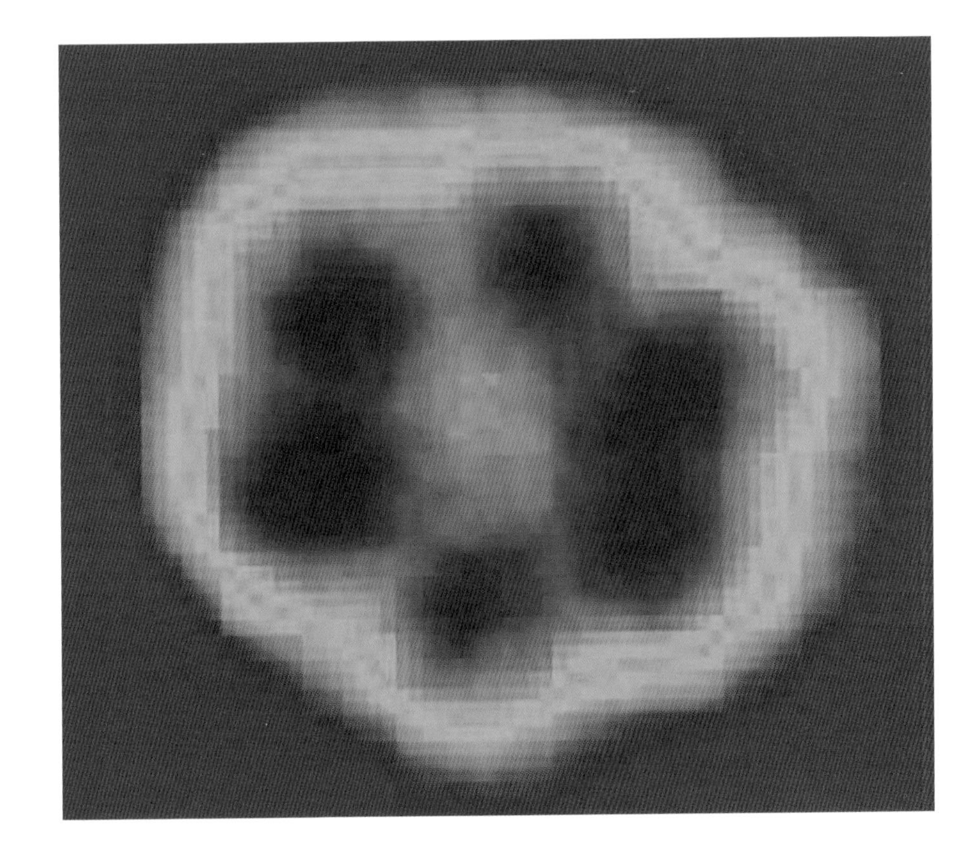

In 2007, a joint project between the London Centre for Nanotechnology, UK, and the Leibniz University in Hannover, Germany, produced this image of a cluster of water molecules (colorized in blue) on a metal surface (red). They described the cluster as the "smallest piece of ice," since the hexagonal shape of the cluster is akin to the structure of ice crystals. *Source:* Angelos Michaelides and Karina Morgenstern, "Ice Nanoclusters at Hydrophobic Metal Surfaces," *Nature Materials* (2007). https://doi.org/10.1038/nmat1940.

Water molecules can also be studied by their interactions with electromagnetic radiation. The positions of water molecules in ice crystals, for example, has been known for more than a hundred years, thanks to X-ray diffraction—a technique in which X-rays are diverted (diffracted) by atoms and molecules. X-ray crystallography is best suited to solids with regular repeating patterns of atoms and molecules, such as ice. But the same approach can be used to show that even liquid water has a "structure," as we shall see in the next chapter. James Watt, Joseph Priestley, Henry Cavendish and Antoine Lavoisier would have been fascinated.

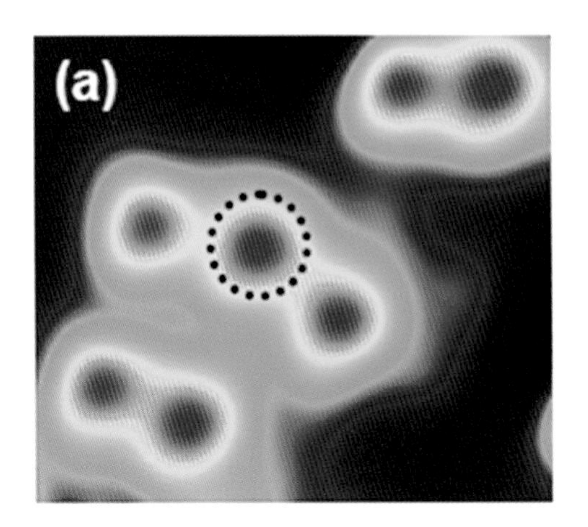

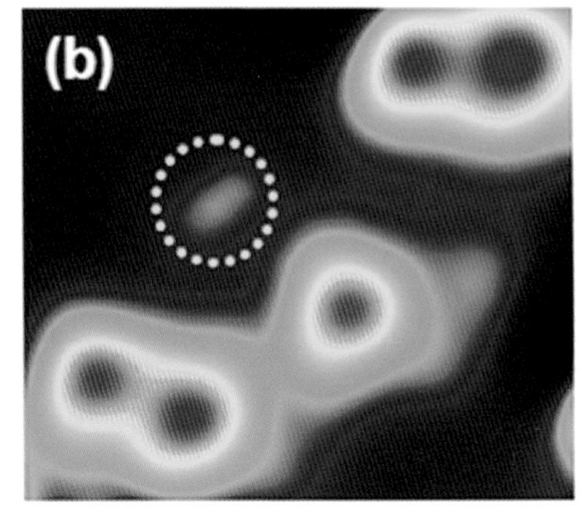

In 2010, a team of researchers at the RIKEN Advanced Science Institute in Wako, Japan, used an STM to interact directly with individual water molecules. They even managed to deliver electric currents, oscillating at the vibration frequencies of the molecules' covalent bonds, to dissociate an individual molecule (in the dotted circle, left) into a hydrogen ion and a hydroxide ion (in the dotted circle, right). *Source:* Hyung Joon Chin, Jaehoon Jung, Kenta Motobayashi, Susumu Yanagisawa, Yoshitada Morikawa, Yousoo Kim and Maki Kawai, "State-Selective Dissociation of a Single Water Molecule on an Ultrathin MgO Film," *Nature Materials* (2010). https://doi.org/10.1038/nmat2740.

CHAPTER 4

Across Three States

In his 1929 poem "The Third Thing," English poet and author D. H. Lawrence wrote: "'Water is H_2O, hydrogen two parts, oxygen one, but there is also a third thing, that makes it water, and nobody knows what it is." Water is so familiar and has been studied so extensively that one might think it is extremely well understood, and as a vapor, it is. Ice, too, has largely been conquered, although scientists occasionally discover or make new forms of ice—more than twenty kinds are known at present. But liquid water is much more of a challenge, and, despite decades of ingenious theory and experimentation, it still defies proper understanding. Hydrogen bonding—that electrostatic attraction that causes water molecules to stick together—seems to be at the root of water's many deviations from normal behavior. Indeed, if pressed, most experts on water would perhaps say that hydrogen bonding is Lawrence's "third thing." And yet while all modern scientific models of the structure of liquid water take into account the hydrogen bonding, not one has so far been able to explain all of water's strange behaviors.

Solid, Liquid, Gas (or Vapor?)

Not all substances can exist in all three states: there is no such thing as liquid wood, for example, and polyethylene has a melting point but no boiling point (since it dissociates into smaller molecules at high temperatures). But every pure element, and many pure compounds, can exist as a solid, a liquid and a gas—or a vapor. Water is the only substance that can exist in all three states at the range of temperatures and pressures found here on Earth. (It is interesting to note that, unlike liquid water and ice, water vapor can never really be in its pure state here on Earth: it is always mixed with other gases, in the air.)

Gas and vapor are very similar: both consist of freely flying particles that are separated by huge distances (relative to their minuscule size). But there is a difference—and whether a substance in its gaseous phase is a gas or a vapor is determined by its temperature and the ambient pressure.

Temperature is related to the kinetic energy of the particles (atoms, molecules, or ions) of a substance. Kinetic energy is the energy of movement; so if the particles of a substance are made to move or vibrate more rapidly, the substance's temperature increases. An individual particle does not have a temperature, since temperature is a bulk measurement—the average over a large number of particles. Pressure, as the name suggests, is a measure of how strongly a surface is being pressed. Rest a standard house brick on the ground face down, and its weight (4.5 pounds) is spread across a relatively large area (29 square inches); the pressure is about 0.16 pounds per square inch. Stand it on one end, and that same weight is concentrated in a smaller area (just over 8 square inches); the pressure on the ground is now more than three times as great, at 0.55 pounds per square inch. Liquids and gases can exert pressure, too—and since they can flow (they are both *fluids*), they press on any surface they meet—not just vertically downward as for the house brick, but sideways or even upward. The weight of the atmosphere presses on all sides of your body, for example. At sea level, the average pressure exerted by the atmosphere is 14.7 pounds per square

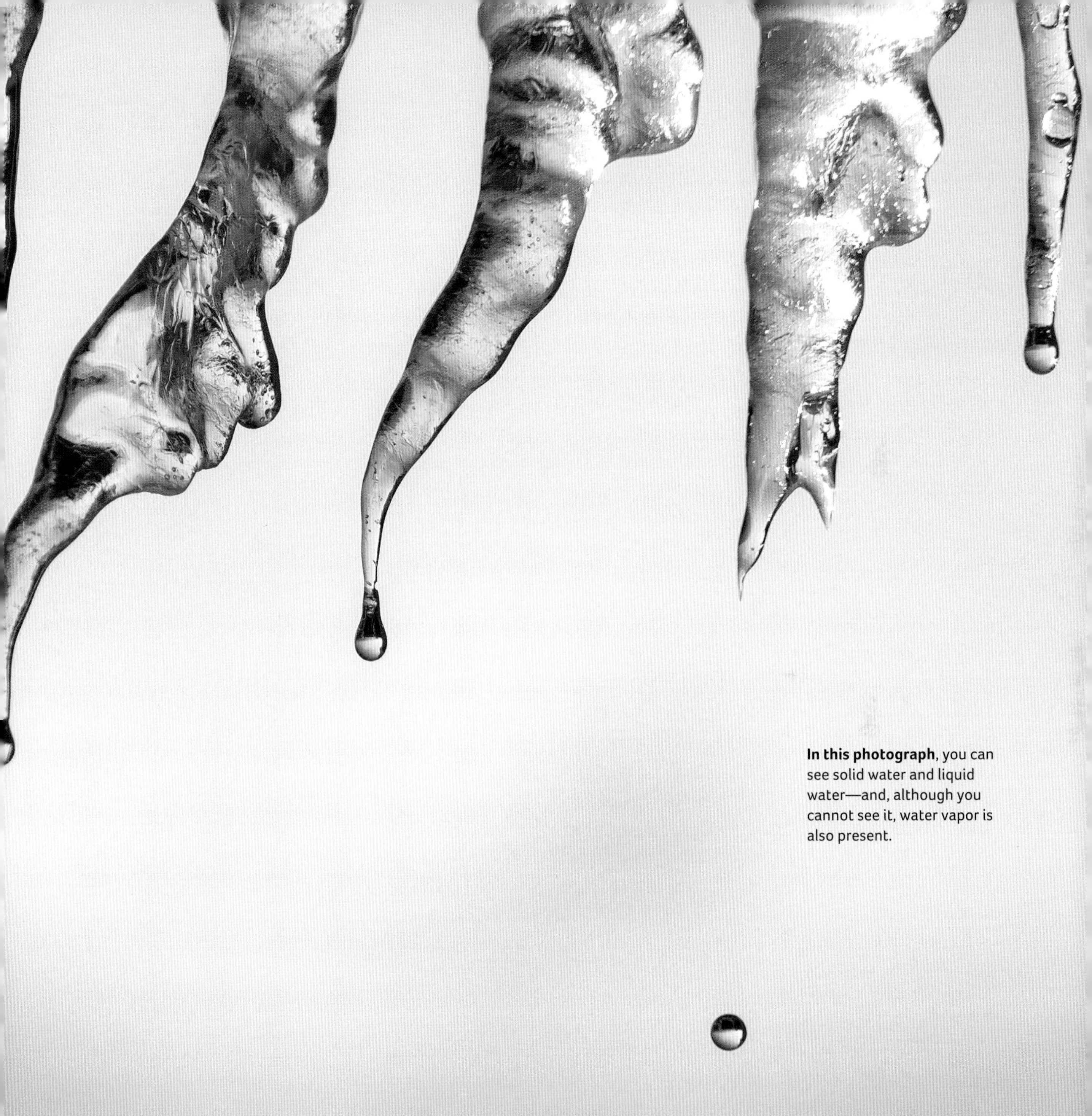

In this photograph, you can see solid water and liquid water—and, although you cannot see it, water vapor is also present.

inch—the same pressure as that exerted by a tower of about twenty-five house bricks standing on their ends.

To understand how temperature and pressure relate to the difference between a gas and a vapor, consider boiling water in a kettle. Boiling is different from evaporation. As noted in chapter 2, some molecules at the surface of a liquid will always have enough energy to escape, so evaporation is always happening. The vapor above a liquid surface has an associated pressure, just as any gas or vapor does. This *vapor pressure* increases with temperature—and when the vapor pressure becomes equal to the ambient pressure, the liquid boils.

At 212°F, the vapor pressure of water is equal to the standard atmospheric pressure of 14.7 pounds per square inch, or 1 *atmosphere.* Atmospheric pressure decreases with altitude or elevation, since the weight of air pressing down on you decreases the higher you go. As a result, the boiling point is lower above sea level. For example, Aspen, Colorado, sits at an elevation of 8,000 feet. The average atmospheric pressure at that altitude is only 0.74 atmospheres—and water boils at about 198°F. At the summit of Mount Everest, where atmospheric pressure is just 0.33 atmosphere, water boils at about 155°F. Meanwhile, if you heat water inside a highly pressurized room, it would boil at a temperature higher than 212°F. This principle is put to good use in a pressure cooker, which cooks foods faster because the temperature inside can rise above the normal boiling point. The pressure inside a pressure cooker is typically around 2 atmospheres, the temperature as high as 250°F. It is informative to plot all of these points on a graph of pressure versus temperature, called a *phase diagram.*

In the phase diagram shown in the figure, there is a line between the liquid phase and the vapor phase—a line that represents how the boiling point varies across a range of pressures. All of the examples in the preceding paragraph are shown as points along that line. It is clear from the diagram that in each case, you could change the vapor back to a liquid just by increasing the pressure—effectively moving vertically upward from those points on the diagram (or, of course, reducing the temperature a little, by moving to the left). Surprisingly, perhaps, the line has an

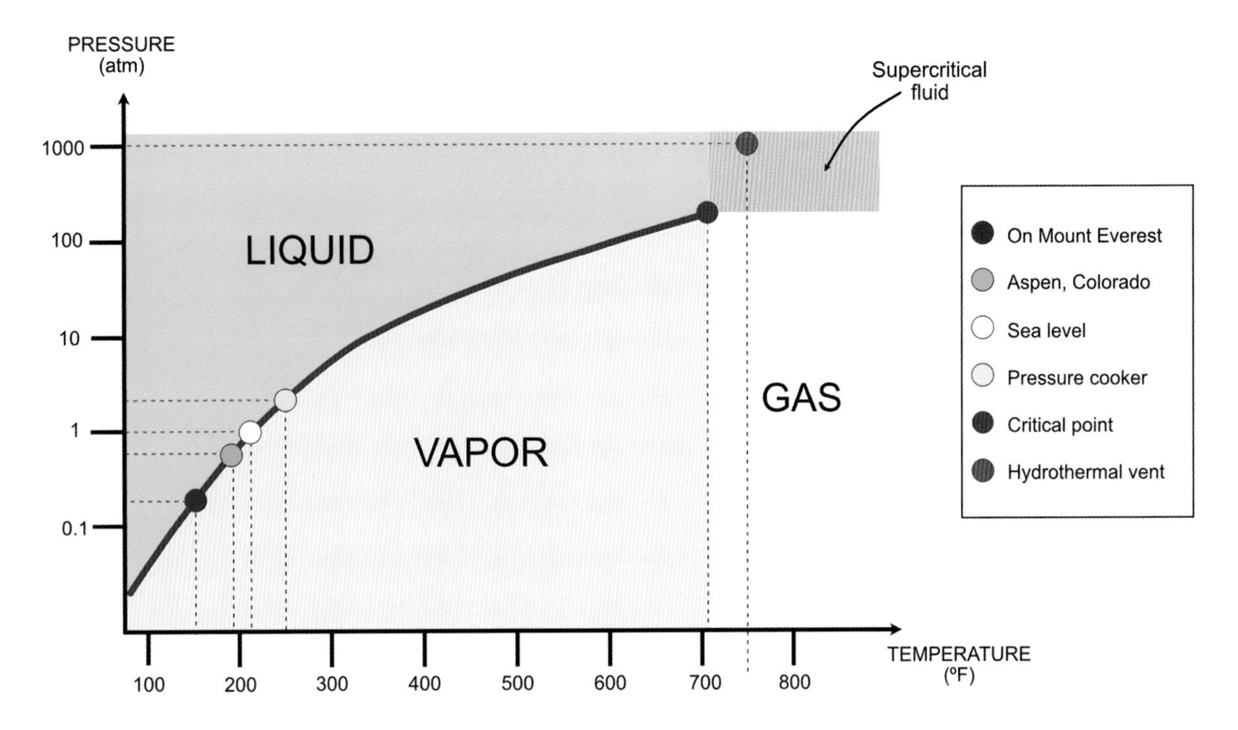

Phase diagram of water for a range of temperature and pressure. Colored dots show points referred to in the text.

end, which is called the *critical point*. For water, the critical point is at 705°F and 218 atmospheres. Above the critical temperature, but below critical pressure, water exists as a gas: it cannot be made to form a liquid just by compressing it, with no change in temperature. And that is the difference between a vapor and a gas: a vapor is a substance in its gaseous phase at a temperature below its critical temperature—one that can be liquefied by pressure alone. That is true for water nearly everywhere on Earth; the critical temperature for oxygen, on the other hand, is –182°F, which is why oxygen is a gas here on Earth. There are some places where the ambient temperature and pressure are both above their critical values for water—where water cannot exist as a liquid or a gas; instead, it is a *supercritical fluid*. One such place is the bottom of the ocean, directly above hydrothermal vents, where molten lava escapes from Earth's crust; temperatures at hydrothermal vents are typically around 750°F and pressures around 1,000 atmospheres.

Making the Change

Pick any point immediately to the left of the line between liquid and vapor on the phase diagram. From there, a tiny increase in temperature (a shift to the right) will cause the liquid to boil. Normally, any heat added to a liquid will increase its temperature, but here, at the boiling point, that is not the case. If you keep heating a pan of boiling water, its temperature will not increase; you cannot heat liquid water to a temperature above its boiling point. Instead, the energy you provide by heating it breaks the hydrogen bonds between molecules, allowing more and more of them to escape the liquid. The energy required to do this is referred to as *latent heat*, and because of the hydrogen bonds holding water molecules together, changing water into vapor requires a lot of it. This is why it takes a long time for a pan left on the stove to boil dry. The amount of latent heat required to change 1 pound of water into vapor—with no change in temperature—is nearly 300 times the amount of heat required to increase the temperature of a pound of liquid water by 1°F. Water vapor carries

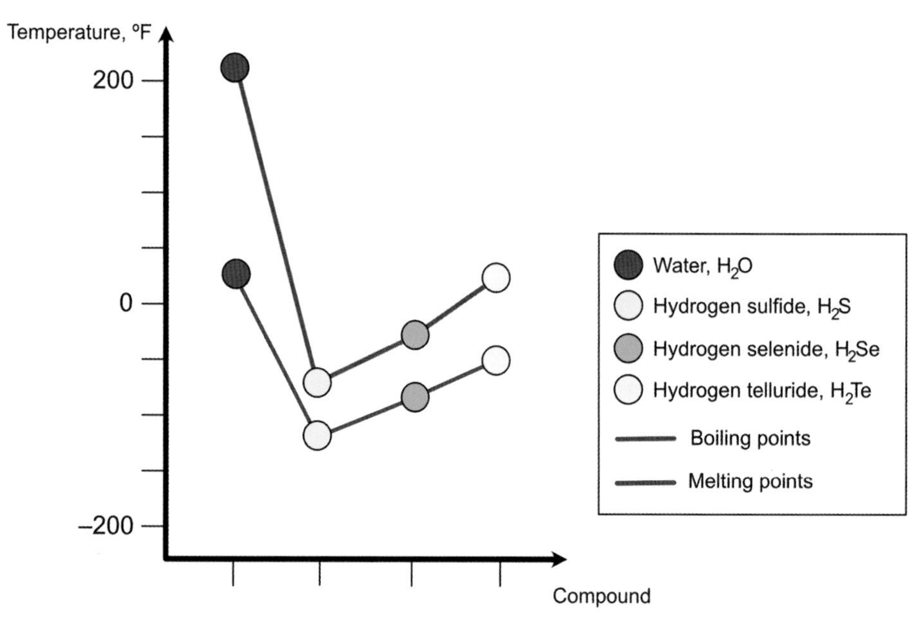

The hydrogen bonding in water is much stronger than in similar compounds. One striking example of this is shown in this graph, which shows the boiling points of compounds H_2X, where X is oxygen (O), sulfur (S), selenium (Se) or tellurium (Te)—all elements in the same group of the periodic table.

this latent heat—whether the vapor was produced by evaporation or by boiling—and when it condenses, it releases it. As a result, steam escaping a kettle can be significantly more scalding than the same amount of liquid water at boiling point would be—and clouds carry and release huge amounts of heat.

Heating water at its boiling point will not increase its temperature—instead, the heat breaks bonds between the molecules, so that they become free of each other.

Any distinct state in which a particular substance can exist is called a *phase* (hence "phase diagram")—and so a change from liquid to vapor is a phase transition or phase change. Extending the phase diagram to include lower temperatures and pressures reveals more phase changes: one between liquid and solid and one between vapor and solid. In each case, latent heat is required or released—depending on which way you cross the line—as before. At the point where the lines meet, called the *triple point*, all three phases of water coexist in equilibrium. Water's triple point, where liquid water is boiling and freezing at the same time, is at a temperature of 32.01°F (just above the normal freezing point) and 0.006 atmospheres (much less than normal atmospheric pressure).

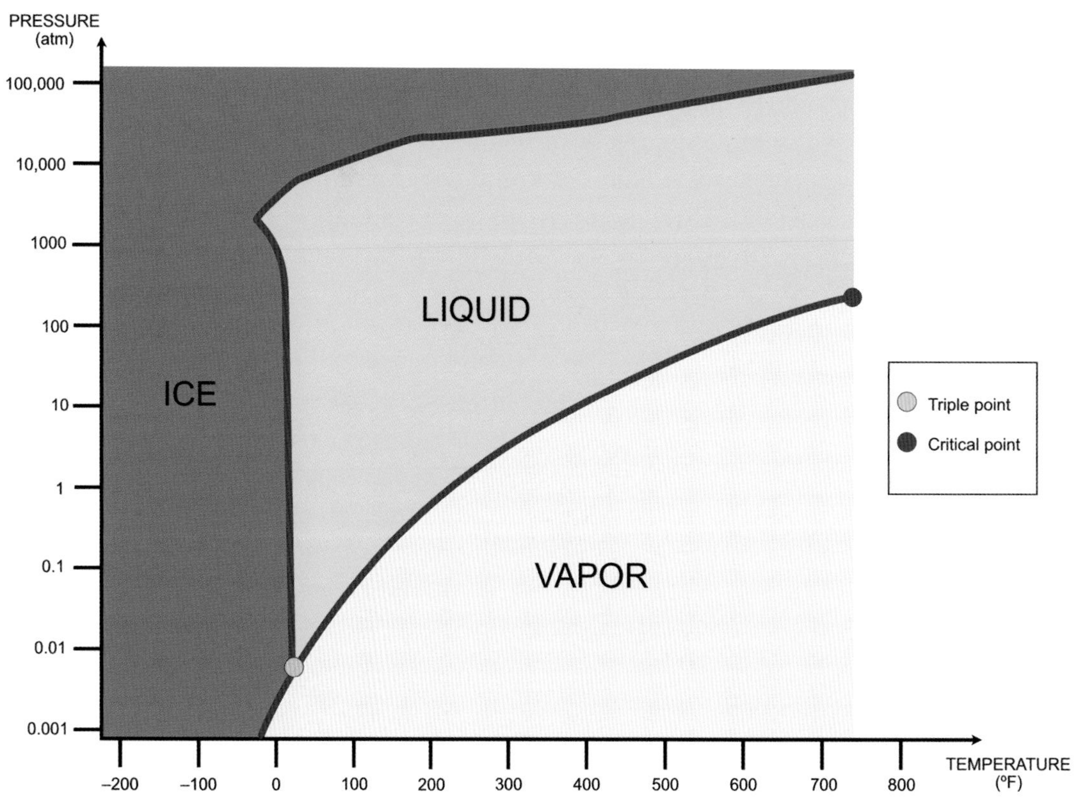

PRESSURE
(atm)
100,000
10,000
1000
100
10
1
0.1
0.01
0.001
ICE
LIQUID
VAPOR
Triple point
Critical point
–200
–100
0
100
200
300
400
500
600
700
800
TEMPERATURE
(°F)

A Solid Foundation

As liquid water cools to its freezing point, its molecules slow down enough that hydrogen bonds can hold them in place. They still have kinetic energy—but it is only vibrational, not translational energy, since the molecules are no longer free to move around. The water molecules fit together in a tetrahedral arrangement, thanks to the near-tetrahedral shape defined by the covalent bonds (O-H) and hydrogen bonds (O:H)—see chapter 3. Each hydrogen bond sits between a hydrogen atom on one water molecule and a lone pair of electrons on the oxygen atom of another. Ordinary ice—the kind that forms at the temperatures and pressures found on Earth—has six-fold symmetry—and it is the tetrahedral pattern that gives rise to this symmetry, as you can see in the image below.

The tetrahedral arrangement of water molecules in ice leads to relatively large open spaces within the crystal structure. It is for this reason that water expands when it forms ice—a phenomenon that people whose pipes have burst in the winter, or who

Inside a bell jar at a very low pressure of 0.006 atmospheres and cooled to 32.01°F—the triple point—water exists in an equilibrium between ice, liquid and vapor. Glass marbles were added to reduce the amount of splashing as the water boiled.

Extending the phase diagram of water over a wider range of temperature and pressure reveals more phase changes and the triple point.

Water molecules held in a tetrahedral arrangement by hydrogen bonding, as found in ice's crystal structure.

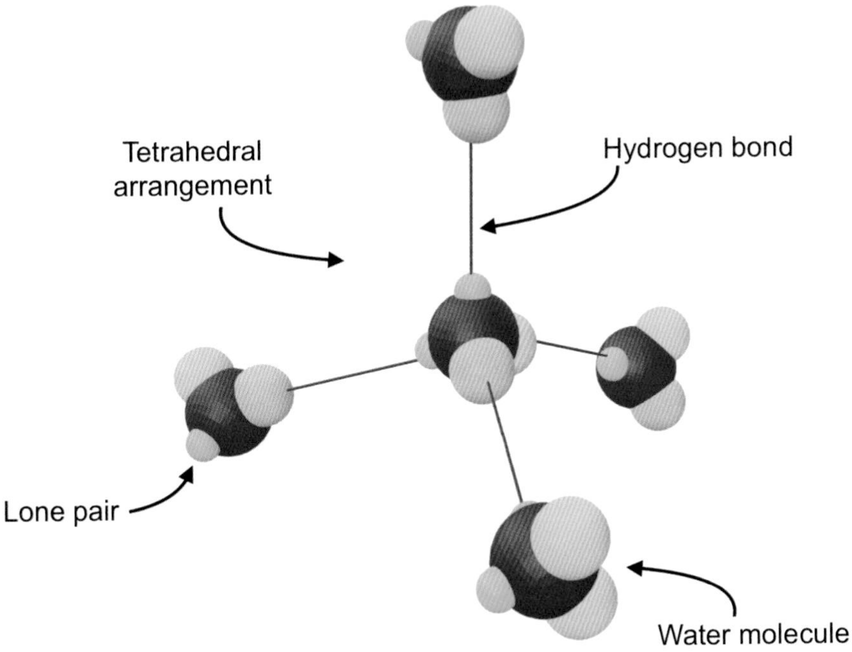

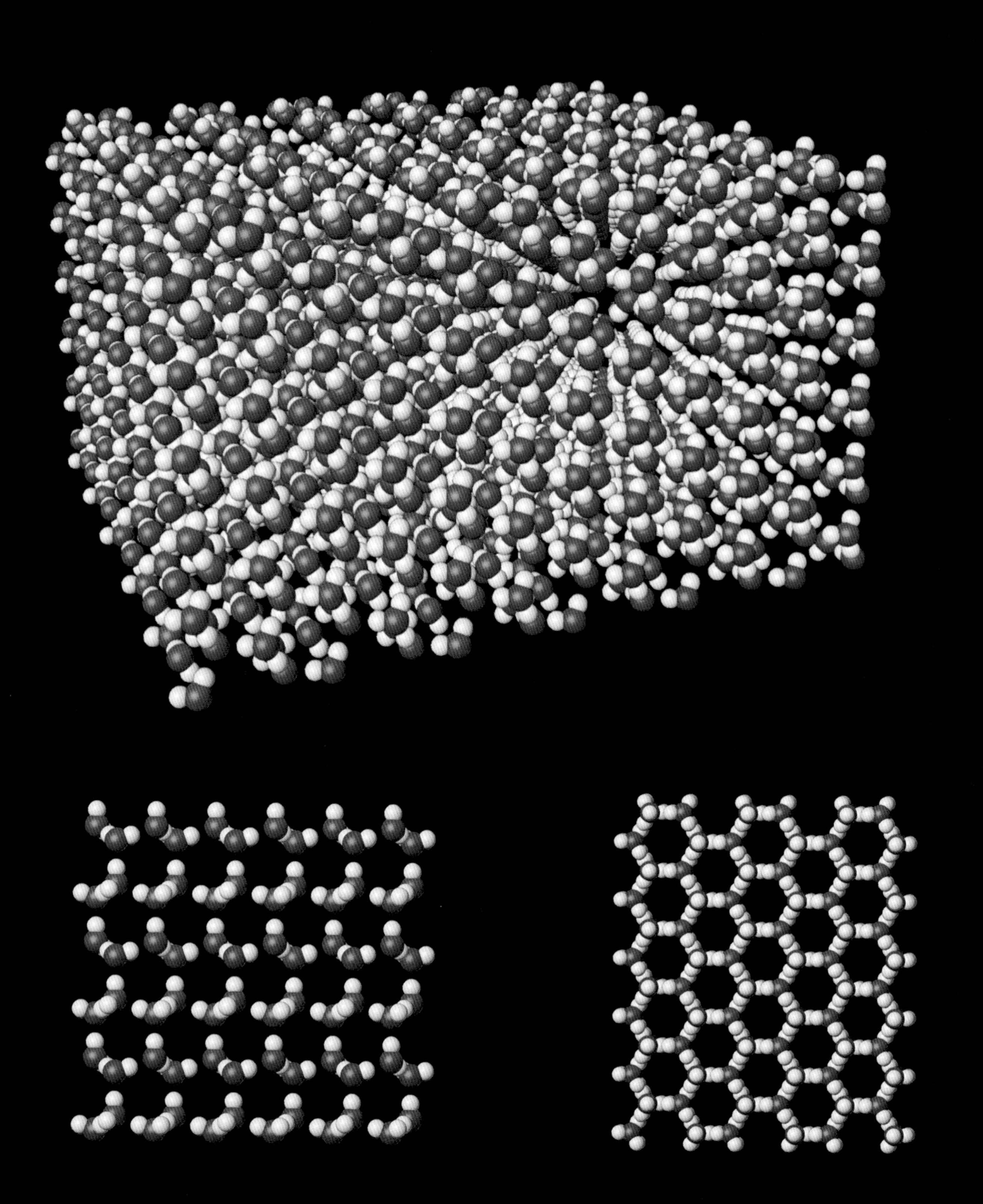

have left a bottle of beer or wine in the freezer for too long, know all too well. Another consequence of the open structure of ice is the fact that ice is less dense than liquid water. This fact—that the liquid is denser than the solid—is almost unique to water. It is of crucial importance to life on Earth, because it means that ice floats. When lakes or rivers freeze, they do so from the top down, so aquatic plants and animals can survive underneath the ice. If ice were more dense than liquid water, it would sink, and more ice would form at the surface, so that bodies of water would freeze solid—and aquatic plants and animals would not be able to survive. Water is most dense at 39.2°F—more than 7°F above freezing, suitable for fish and other animals that live in water in cold climates. It is worth noting the fact that water at that temperature is at a density maximum: it will expand whether you heat it or cool it.

Three different views of ordinary ice, or "ice-Ih" (the h is for "hexagonal"). Notice the large open spaces between the water molecules, and the hexagonal symmetry—both features of the tetrahedral pattern required by the hydrogen bonding.

When liquid water cools to its freezing point, its molecules arrange into an open crystal structure, resulting in a significant expansion of its volume. This can lead to some unfortunate effects.

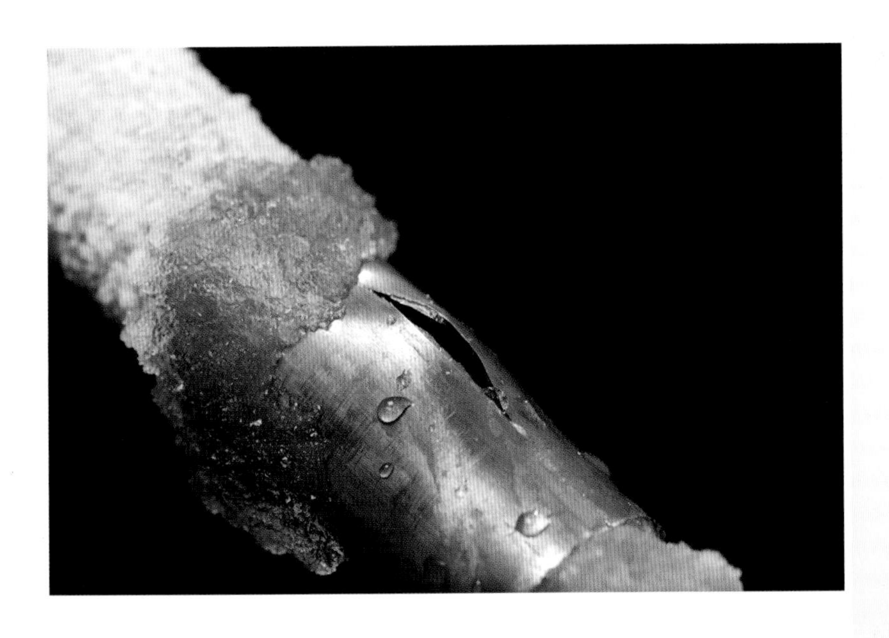

Another consequence of ice's open crystal structure is that putting pressure on ice can cause it to melt. This is why the near-vertical line between liquid and ice on the phase diagram slopes up and slightly to the left. For nearly all other known substances, that line slopes up and to the right. Pressure melting of ice is often used as an explanation of why ice is so slippery. The story goes that the pressure exerted by an ice skate is quite large and melts the ice, forming a lubricating liquid layer that allows the skate to glide along freely. But this is not the case: the pressure is not nearly enough to melt the ice, even under the narrow blade of a skate—and ice is no less slippery on ordinary shoes. The real reason concerns a liquid-like layer at the surface of the ice, as explored in chapter 5.

Burning Ice

◀ **Life can persist** even when ambient temperatures above the surface are below freezing, thanks to the fact that ice floats.

Yet another consequence of ice's open crystal structure is that the cavities within the array of water molecules can be home to small molecules of other compounds, such as methane—which is normally found as a gas in underground oil reserves. When molecules of methane become trapped in water ice, the arrangement of the water molecules shifts into one of several different cage-like, or *clathrate*, crystal structures—and the resulting mixture is called a *methane hydrate*.[1] When a certain minimum number of the cavities become occupied by methane molecules, the whole mixture becomes stable—even at temperatures several degrees above ice's normal melting point. The amount of methane trapped inside these hydrates can be surprisingly large: the effective density of the methane present is about 160 times as great as methane's density in its gaseous form.

Methane hydrates form at relatively high pressures, and temperatures close to water's normal freezing point—conditions commonly found at the seafloor at some coastal margins. Most of the methane hydrates around the world today contain methane that was produced by organic matter that decayed underwater after the sea level rose at the beginning of the current warm interglacial period. Geologists estimate that the amount of methane locked away in methane hydrates under the oceans is several times more than all the known methane gas reserves. To some, this presents an opportunity: mining the seafloor for methane hydrates could provide a new, plentiful source of fuel. However, that would run counter to current efforts to reduce carbon emissions, which is part of humanity's attempt to slow down climate change. Indeed, many climatologists worry about the potential impact of methane hydrates even without anyone mining them and burning the methane. With such vast reserves of methane locked away in them, the melting of even a tiny proportion of the world's methane hydrate inventory would release huge amounts of methane into the atmosphere—and methane is a far more potent greenhouse gas than carbon dioxide. A small proportion of methane hydrates is found in permafrost inside the Arctic Circle, where some have already begun melting, as the climate has warmed over the past several decades. On a more positive

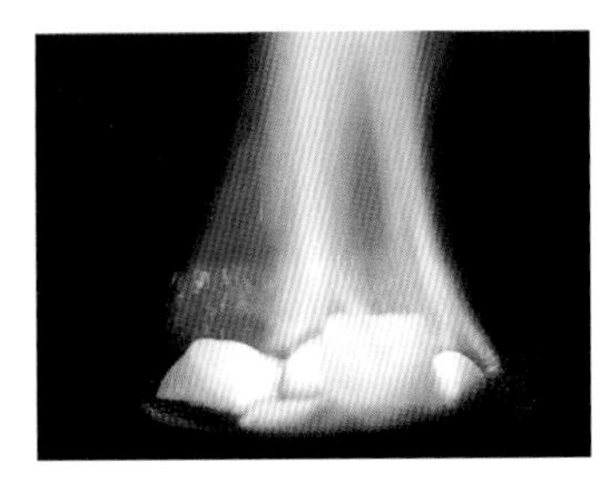

▲ **Methane hydrate** burning. *Source:* USGS.

▼ **The white, icy material** beneath the rock at the center of this photograph is methane hydrate. The photograph was taken in the Gulf of Mexico, at a depth of about 3,170 feet. *Source:* USGS.

slant, methane hydrates might also help in the fight *against* climate change. Methane is not the only small molecule that can be trapped in water's crystal structure: several research projects are underway that aim to pump carbon dioxide produced in the burning of fossil fuels into water ice for long-term storage, or sequestering.[2]

It's Just a Phase

The familiar type of ice—the ice cubes in a freezer, snowflakes that land on a baseball hat, and glaciers from top to bottom—makes up nearly all the solid water found naturally on Earth. But it is not the only kind of solid pure water—and in the Universe at large, it is not even the most common. It is just one of several distinct phases, each with a characteristic and unique set of properties, such as density and heat capacity, and a different arrangement of its constituent water molecules. As a result, many other phase-change lines can be added to the phase diagram, criss-crossing the darker blue region, in which ice exists. (Notice how, on that diagram, the "ice" region extends out to surprisingly high temperatures and extreme pressures.) "Ordinary ice" gained the title "ice-I" in 1900. In that year, German chemist Gustav Tammann subjected ice to pressures of a few thousand atmospheres and saw it change, twice, into two distinct phases, which he named ice-II and ice-III.[3] Then, in 1942, ice-I had to be renamed again, when German crystallographer Hans König observed that a slightly different form of ice-I forms at very low temperatures.[4] König's new phase has a "cubic" crystal structure (while maintaining the tetrahedral bonding pattern), and so it became "ice-Ic" (cubic), while the familiar phase became "ice-Ih" (hexagonal). Ice-Ic forms at ordinary or low pressures, and temperatures of between –170 and –22°F in freezing water droplets[5]—meteorologists suspect that it exists in high altitude clouds, probably with ice-Ih in a "stacking disordered" mixed crystal.[6]

One scientist more than any other pioneered the investigation of the phase diagram of ice, particularly at high pressures:

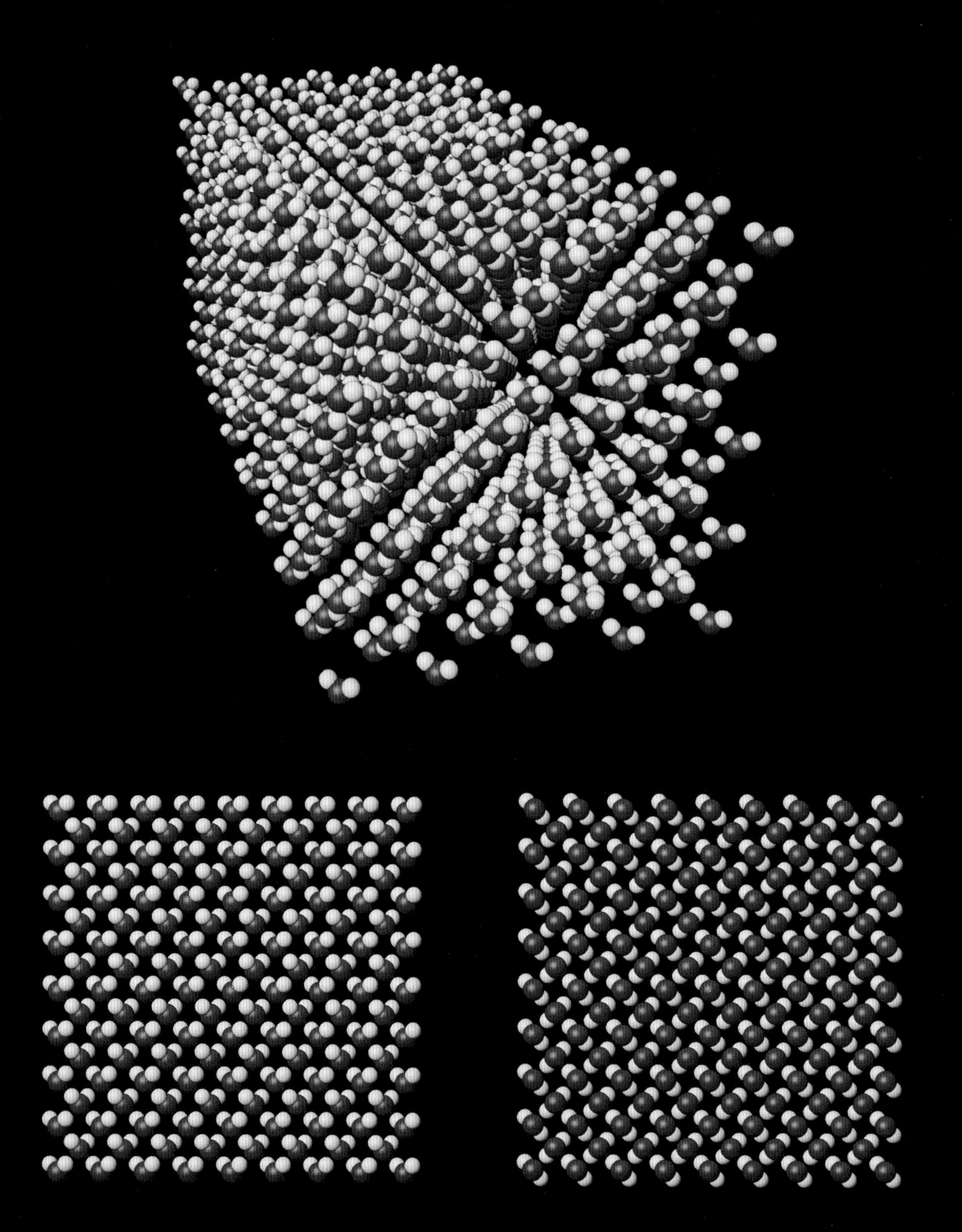

Three different views of "ice-Ic" (the c is for "cubic"). Compare this with the structure of ice-Ih on page 112. All the water molecules are joined in a tetrahedral pattern in this crystal structure.

American physicist Percy Bridgman. In 1910, he invented a device that could produce pressures of around 75 tons per square inch (about 10,000 atmospheres).[7] Later versions could develop as much as 700 tons per square inch (nearly 100,000 atmospheres). Bridgman discovered several high-pressure phases of ice. Under pressure, the hydrogen bonds between water molecules and the covalent bonds within them can become distorted. In *superionic* ice phases, the covalent bonds break, and the crystal structure is then made of ions rather than covalently bonded atoms. In the superionic ice featured in chapter 1, which is known as ice-XVIII, the ions are oxygen (O^-) and hydrogen (H^+), but some superionic ices involve different combinations of ions. Superionic ice phases are no longer composed of H_2O molecules, but they are *stoichiometrically* still water: hydrogen still outnumbers oxygen precisely in the ratio 2:1.

Percy Bridgman at Harvard University, working with his 1938 press, which could exert pressures of 55 tons per square inch. *Source:* Emilio Segre Visual Archives.

At the time of writing, twenty different crystalline phases of ice have been discovered.[8] Apart from the superionic forms, where the molecules are no longer intact, all the crystalline phases follow a set of parameters, called the *ice rules*, which were determined in the 1930s. According to these rules, each oxygen atom is surrounded by four other oxygens, with a hydrogen along the line that joins them. Two of those hydrogens are bonded covalently, and are part of the same water molecule as the oxygen in question, while the other two belong to neighboring water molecules, and are held in place by hydrogen bonds. Despite these strict rules, theoretical studies suggest that around 300 different crystalline phases of solid water may be possible altogether, including some that are far less dense than ordinary ice.[9]

As well as the crystalline phases, solid water can be *amorphous*. The archetypal amorphous substance is glass, which is made of silicon dioxide. When molten silicon dioxide cools quickly, the silicon and oxygen atoms do not have time to assume positions in a regular array; unable to move or even rotate, the atoms are frozen in disarray. The water molecules in amorphous ices are similarly disordered. Amorphous ice has been detected in large quantities in interstellar space, in the giant molecular gas clouds in which stars form (see chapter 1). It was first produced in the laboratory in the 1930s, by cooling water vapor to low temperatures extremely quickly. Study of this "glassy" ice with X-rays found no regular structure.[10] In 1984, researchers made a new form of amorphous ice, by using high pressure to melt ordinary ice at very low temperature (–322°F) and then warming it, rather than cooling very quickly.[11] A year later, the same researchers found they could make it by compressing the original kind of amorphous ice at a low temperature. Since this new form has a higher density than the other one, the two phases became known as low-density and high-density amorphous ice (LDA and HDA). In 1996, a very high density phase (VHDA) was discovered. A recent supercomputer simulation of the transition between low-density and high-density amorphous ice suggested that there is more order in these forms of ice than has long been

These photographs of low-density (left) and high-density amorphous ice come from the National Institute for Materials, Japan.

assumed.[12] Since amorphous water is like a snapshot of liquid water, molecules frozen in time, the order detected in the simulation might be due to order in liquid water—more of which below.

Diving into Liquid Water

A glass of water may look *homogeneous* (the same throughout), but there is structure to it (it is *heterogeneous*). So ubiquitous and important is water that determining the structure of liquid water, and working out how the structure relates to water's properties, is of great importance in engineering, biology, geology, chemistry, climate modeling, astrophysics and many other fields. Despite that—and despite the fact that it is one of the most closely studied substances—water in its liquid form is poorly understood. In its 125th anniversary edition, the American Association for the Advancement of Science's magazine included the structure of water among 125 "big questions that face scientific inquiry over the next quarter-century."[13]

Even simple liquids are generally more difficult to understand and model than gases or solids. Gases (and vapors) are predictable, precisely because they have no "structure" or "order" whatsoever. Each particle is fast moving and free-flying, with a random speed and direction of movement. Despite this randomness, there exists an entirely predictable distribution of speeds at a given temperature, and gases are well understood. Solids are predictable and well understood for the opposite reason: because they possess a regular structure throughout (*long-range order*), thanks to the repeating pattern of their crystal structures. Liquids fall between the two extremes: their particles are free to move around, like those in a gas, but they do have some *short-range order*, some structure to them. To see why, consider a simple liquid, made of individual atoms. The atoms can be as close to each other as their van der Waals radii will allow (see chapter 3)—so a simple liquid can be pictured as a huge collection of jostling marbles. In such a system, each marble is surrounded by a "shell" of nearest neighbors, and that is surrounded by a looser

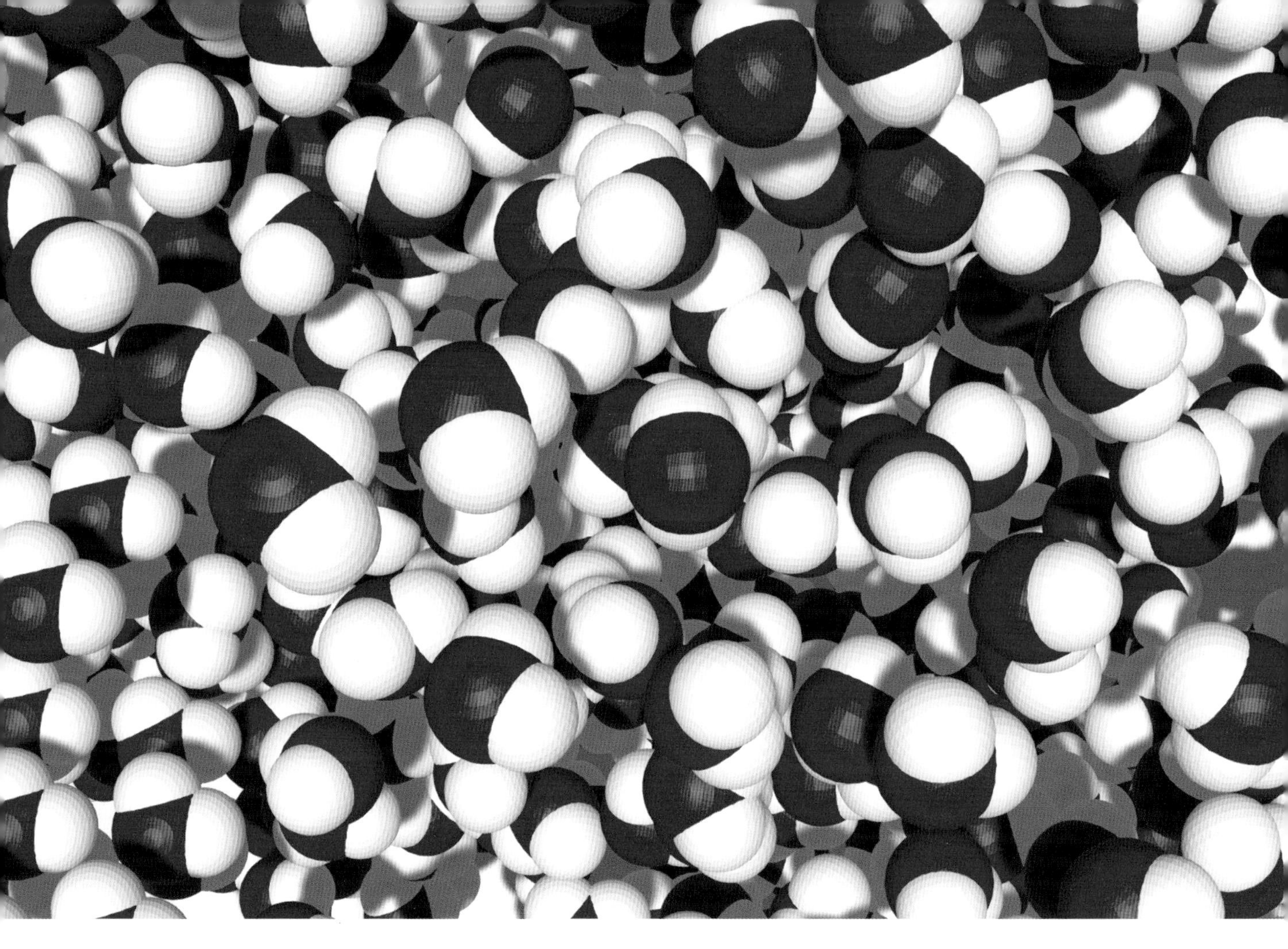

second shell. The existence of these shells provides the local, short-range, order—but beyond, the particles are more randomly arranged. And so, a liquid lacks the predictable long-range order of crystalline solids, and the predictable lack of order of a gas.

The differences in structure between solids, liquids and gases can be illustrated using a *radial distribution function* (RDF). This is a mathematical tool that plots the probability of finding particles within a substance, as a function of the distance from any particular particle. For a particle in a gas, there is an equal probability of another particle being found at any distance. The RDF for a gas is therefore a horizontal line, the probability remaining the same at all distances. In a crystalline solid, there will be certain distances where there is zero chance of finding

▲ **A basic model** of what liquid water is like at the molecular scale. In this picture, the molecules are arranged randomly, and also randomly oriented. In reality, hydrogen bonds give some structure to liquid water—but no one can work out exactly what that structure might be.

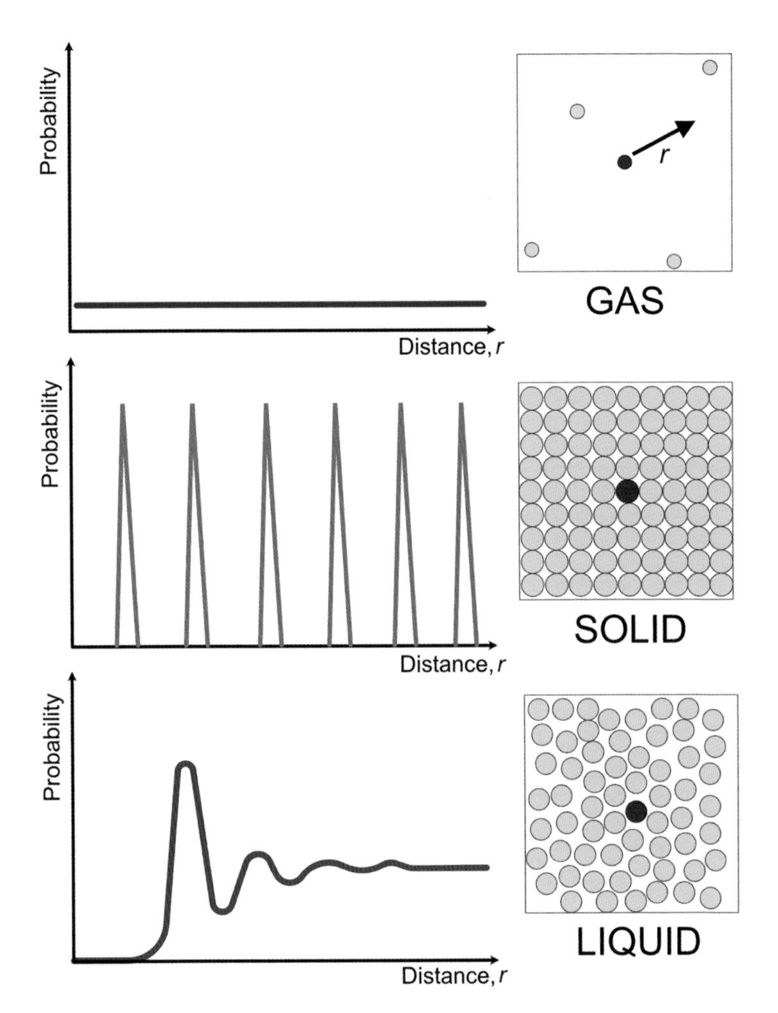

▲ **Simplified radial distribution functions** for a typical gas, solid and liquid. Note that the RDF curve for the gas is on a different distance scale, since the average distance between any two particles in a gas is far greater than in a solid or a liquid. The horizontal axis represents distance, *r*, from a given particle (shown in purple). The vertical axis is the probability that another particle will exist at distance *r*.

a particle (consider the open structure of ice, for example), but a very high probability at very particular distances, thanks to the repeating crystal pattern. In a liquid, the RDF shows a high probability of particles being found at the distance of the first shell, a smaller probability for particles in the second, and perhaps a third—but beyond that, the RDF tails off to a constant value, a straight line as seen in the RDF of a gas, reflecting the lack of long-range order. Importantly, the undulations of an RDF curve for a liquid are less sharp than in a solid, because the particles are jostling around, and so, even at close range, have a randomness to their positions.

For liquids whose particles are molecules, rather than individual atoms, there are added complications. Molecules can be nonspherical, and some possess electric charges—these features can disrupt or enhance local order, and water molecules possess both of them. Scientists with a hypothesis about the structure of a substance can work out what the RDF would look like, and then test their hypothesis by measuring the RDF in experiments, which typically involve X-rays or neutrons passing through the substance. Measurements of water's RDF can determine the average oxygen–oxygen distances, hydrogen–hydrogen distances, and hydrogen–oxygen distances. They can therefore provide information about not only the overall arrangement of molecules in water, but also the lengths of the bonds within water molecules (those lengths can change under the influence of other molecules nearby—another source of complication in water's structure). No surprise, then, that the radial distribution function has been a crucial part of the quest to determine the structure of liquid water for nearly a hundred years.[14]

Another tool available for scientists seeking to crack water's code is computer modeling. Most computer models of water are based on *molecular dynamics*, which utilizes the size of the water molecule, based on the van der Waals radius, as well as its charge distribution, to calculate the forces between any two molecules, and so determine how the molecules will move relative to each other. Some of the most complex computer models also make use of subtle quantum mechanical interactions between the electrons and even the nuclei of neighboring molecules. Computer models predict when hydrogen bonds will form, and when they will not. A molecular dynamics simulation involving just two molecules would reproduce the behavior of the particles extremely well. But introducing even one more molecule leads to a level of complexity that can never be modeled with total accuracy. This is an example of the *three body problem*, in which the interactions of three objects introduce chaotic, unpredictable behavior. The three body problem dates back as far as 1687, when Isaac Newton used his equations of motion and gravity to investigate the Sun–Earth–Moon system. Newton realized that trying to work out the mutual influence of all three objects would prove intractable. A computer model that includes hundreds, or even many thousands, of water molecules is therefore definitely intractable—and as a result, computer modelers have to make compromises. Despite these compromises, sophisticated modern computer models provide a unique and vital window into water's behaviors, and a way to test hypotheses against experimental data.

The ultimate test of whether a hypothesis or computer model of liquid water is on the right track is how good it is at explaining or predicting water's anomalous behaviors. Scientists studying water have identified around seventy anomalies,[15] mostly concerning the behaviors of liquid water or amorphous solid water. Many of these anomalies are obscure, and of direct interest only to physicists and chemists. Here are a few examples: the nuclear magnetic resonance spin-lattice relaxation time is very small at low temperatures; the dielectric constant is peculiarly high; the relative permittivity has a temperature maximum; the specific heat is related to entropy fluctuation . . . you get the idea. The best known anomaly of water is its density maximum, at 39.2°F, as described above. Attempts to account for that fact began in the late nineteenth century.

A Cluster of Ideas

In 1892, German physicist Wilhelm Röntgen (who went on to discover X-rays, in 1895) tried to explain water's density maximum anomaly by suggesting that as ice melts, small parcels of its open, low-density structure survive in the more dense liquid water. In other words, cold liquid water might be a mixture of free, closely packed molecules (high density) and small clusters of water molecules, essentially small pieces of the open (less dense) ice structure. As the temperature rises, the ice particles would melt away—and at the density maximum at 39.2°F, there would be only the higher density component left. Increase the temperature still further and thermal expansion would take over—the normal behavior of liquids.

In 1933, Irish crystallographer John Bernal and English physicist Ralph Fowler developed this idea further. Considering the density anomaly, they wrote: "A simple disordered close packed assembly of water molecules of radius 1.4 Å would have a density of 1.84 [grams per cubic centimeter]; or conversely, for a density of 1.00 the equivalent radius would be 1.72 Å. We have therefore the choice of assuming either that water is a simple, close packed liquid in which the effective molecular radius has changed from 1.4 Å in the solid to 1.72 Å in the liquid, or that the radius is still approximately 1.4 Å but that the mutual arrangements of the molecules are far from that of a simple liquid."[16]

In their paper, the two scientists utilized the newly emerging disciplines of X-ray crystallography (normally used to determine the structure of solids) and quantum mechanics, along with the recently discovered notion of the hydrogen bond in water and the proposed open-V shape of the water molecule, to suggest that water molecules arrange themselves tetrahedrally. Bernal and Fowler established the ice rules, but they realized that the tetrahedral bonding occurs not only in ice but also in liquid water. They compared the tetrahedral bonding pattern in water with a similar arrangement of the atoms in quartz and tridymite—two different structures of the compound silicon dioxide. The density of quartz is significantly higher than that of tridymite, and Bernal and Fowler suggested that water molecules might adopt similar dualistic bonding patterns in the liquid. The idea that there are two distinct phases of water in a single liquid is controversial, but it persists in various forms in many current hypotheses about the structure of water.

In the 1940s and 1950s, the idea of water clusters developed, and grew in prominence and popularity. Some hypotheses included a range of sizes of the ice-like clusters of water molecules—dimers (made of two molecules), trimers (three), tetramers, pentamers, hexamers, heptameters and octamers. The amount of space taken up per water molecule increases with the number of molecules present—so the density of water would vary with the proportions of the different clusters present. In 1957, chemists Henry Frank and Wen-Yang Wen, at the University of Pittsburgh, presented a hypothesis involving dynamic, or *flickering*, water clusters.[17]

The first few water clusters—from the dimer (two molecules) up to the octamer (eight). These clusters, and many other arrangements, have been detected and studied in water vapor and in liquid water in confined spaces—and it is very likely that they exist in bulk liquid water, as "flickering" structures, with molecules constantly joining and leaving.

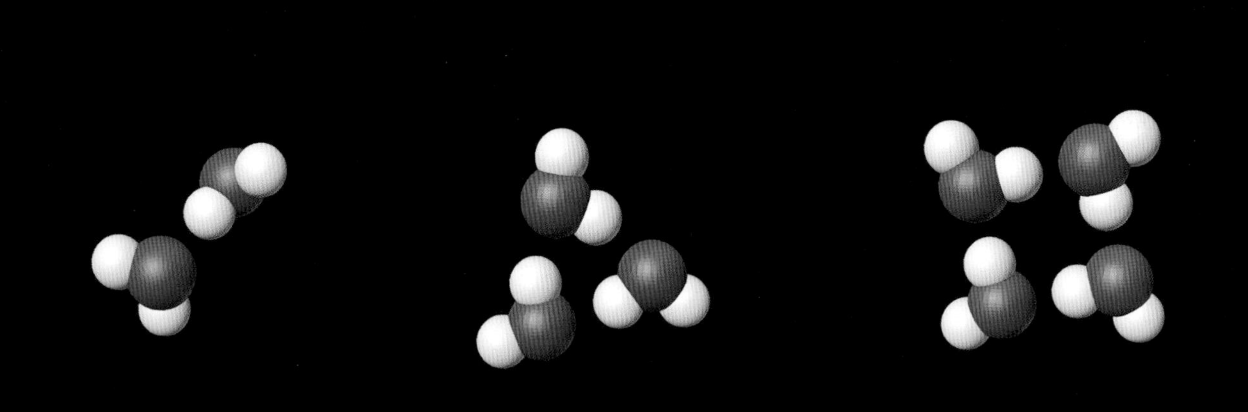

Their model set out to account for several of water's anomalies, not just the one involving water's density maximum—and also to explain how water interacts with dissolved substances (more on that in chapter 5). Frank and Wen-Yang suggested that the clusters would be in constant flux, with lifetimes of trillionths of a second. A variation of the idea of ice-like clusters was proposed in the 1950s and 1960s. In this case, there would be only one fairly open structure, inside which individual water molecules could fit. These *interstitial* water molecules would increase water's density overall—and some current models of liquid water include interstitial water.[18]

The existence of clusters of water molecules in liquid water captured the imagination of some people beyond the scientific community. Since the 1990s, several bottled water products have been introduced onto the market that purport to be "healthier" because they contain predominantly smaller clusters compared with tap water. Despite the unscientific nature of those claims, the idea of flickering clusters remains at the heart of many of the modern models of liquid water. Typically, in these models, the average cluster size and the proportion of clusters present in liquid water decrease as the temperature rises. Researchers using a variety of techniques have found direct evidence of the existence of clusters in liquid water, albeit mostly in tiny droplets or jets,[19] in confined spaces or when other substances are dissolved in water.[20]

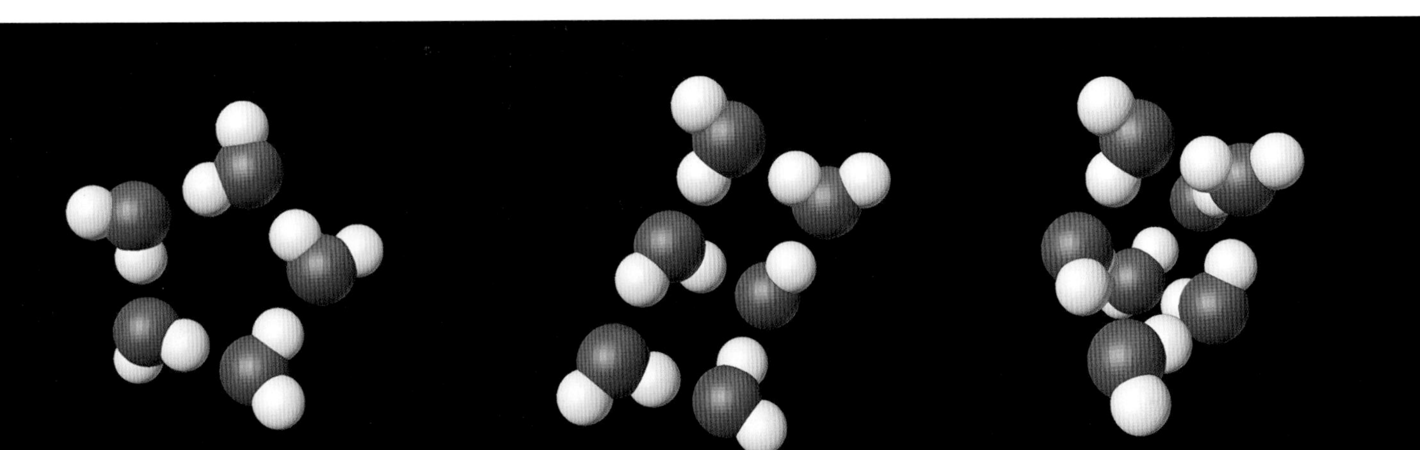

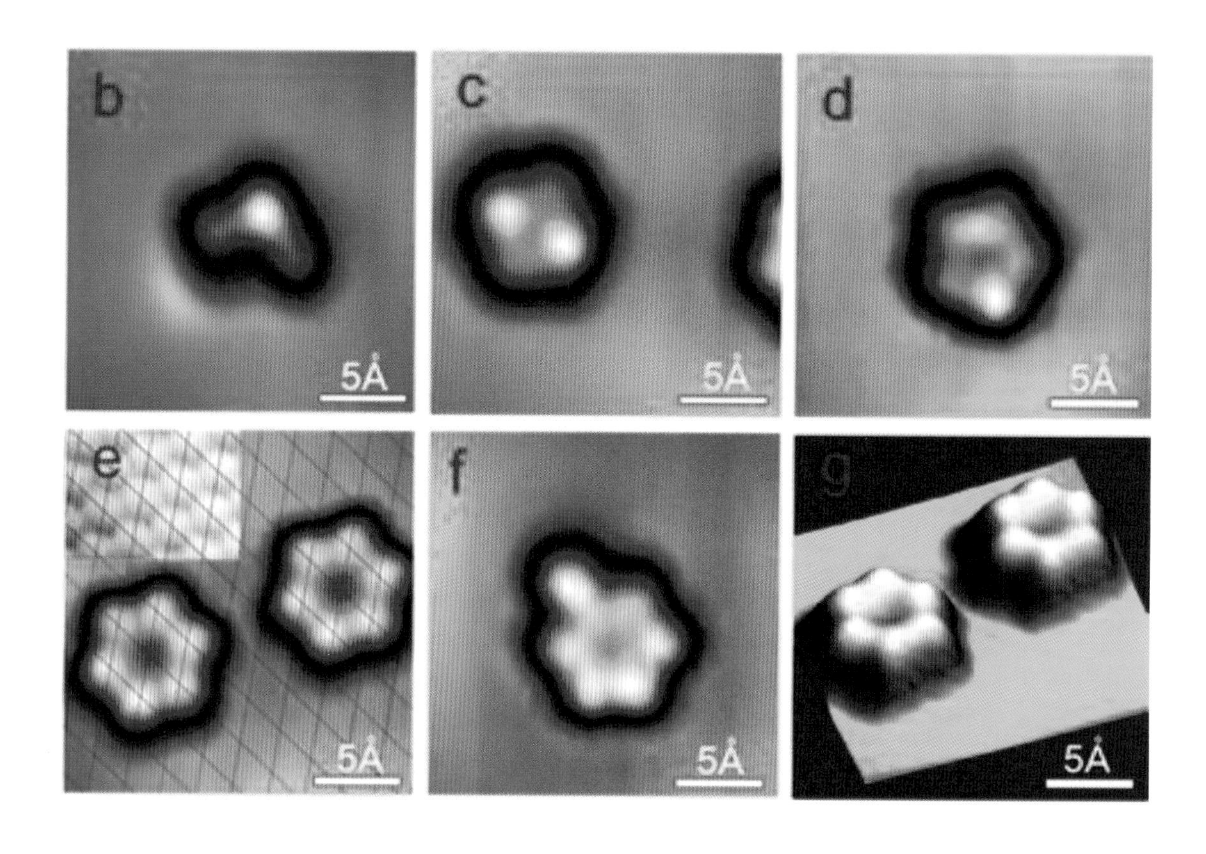
b
c
d
e
f
g
5Å
5Å
5Å
5Å
5Å
5Å

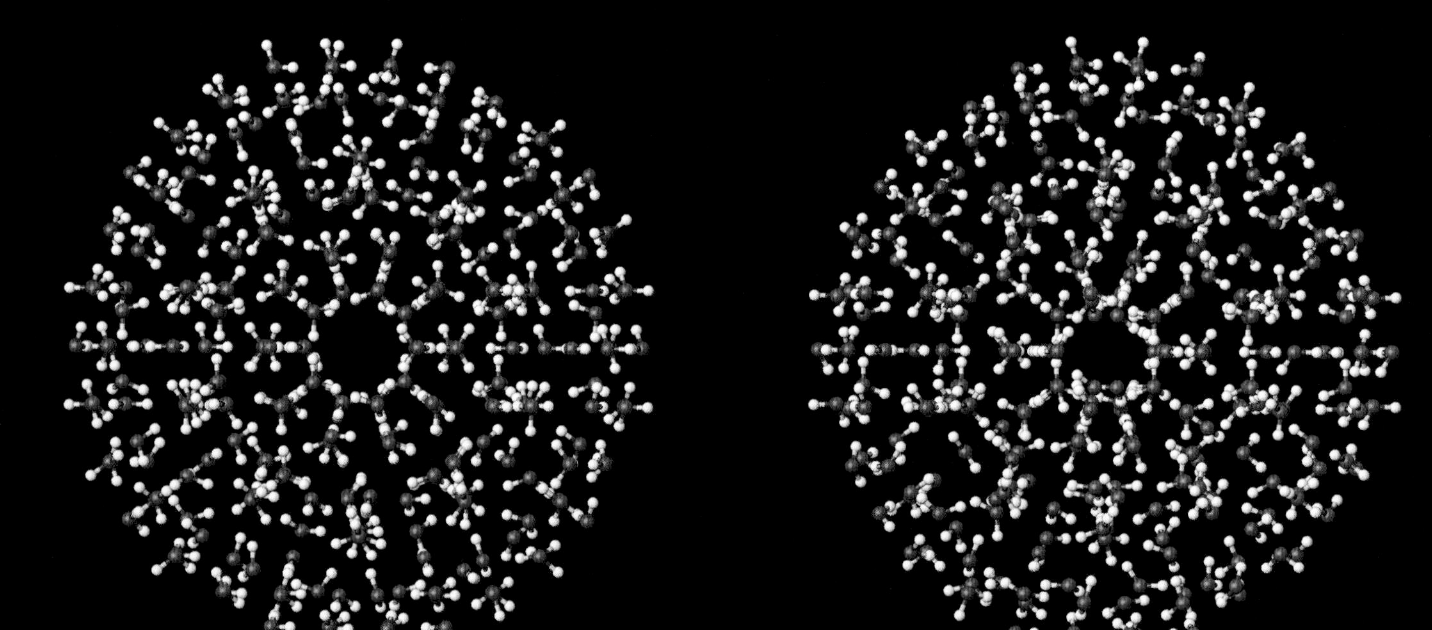

Water clusters on a copper surface, captured by a scanning tunneling microscope in 2018, by a team of scientists in China. Panel (b) shows a "trimer," a three-molecule cluster, while images (c)–(f) show a tetramer, pentamer, hexamer and heptameter, respectively. Panel (g) is a three-dimensional realization of the hexamer. *Source:* Anning Dong, Lei Yan, Lihuan Sun, Shichao Yan, Xinyan Shan, Yang Guo et al., "Identifying Few-Molecule Water Clusters with High Precision on Au(111) Surface," *ACS Nano* (2018). https://doi.org/10.1021/acsnano.8b02264.

Some modern cluster models include much larger assemblies, including the icosahedral model, which includes clusters containing 280 water molecules,[21] as well as smaller clusters. If they really exist, these large clusters would be stable entities, albeit flickering, with water molecules continuously joining and leaving. The dynamic reorganization of the hydrogen-bonded structures in liquid water is on typically on the scale of the picosecond (one trillionth of a second, 10^{-12} second)—this is the length of time a typical hydrogen bond lasts between two water molecules. A hydrogen bond does not stay broken for long: there is a much shorter period of time when two water molecules are not involved in hydrogen bonding. Whatever the precise details of the structures that form in liquid water may turn out to be, it is certain that there exists a vast interconnected network of molecules attracted to one another by fleeting but ever present hydrogen bonds.

Super Cool

In the icosahedral cluster model, formulated by Martin Chaplin, molecules self-assemble into large geometrical clusters, each containing 280 water molecules. There are two forms of these clusters—an expanded and a condensed version—each with a different impact on water's density. In these pictures, for the sake of clarity, the atoms have been shown smaller than the van der Waals radius, and bars represent the covalent bonds. *Source:* Data from Martin Chaplin.

Most of water's anomalies are significant or apparent only in relatively cold water, below about 120°F. Above that temperature, water behaves pretty much like a typical liquid (apart from its high heat capacity and high boiling point). As the temperature drops toward freezing, the average kinetic energy per molecule decreases, allowing hydrogen bonding to play an ever greater role. For this reason, *supercooled water*—water that remains liquid despite its temperature being below the normal freezing point—is of particular interest to researchers. Supercooled water can only be produced from water with very few impurities. As noted in chapter 2, a solid or a liquid surface is normally required in order for ice to *nucleate*, or begin forming; the smallest impurities in the liquid can provide surfaces for nucleation. Nevertheless, careful experimenters can cool very pure water down to around −40°F without it freezing. At higher pressures, liquid water can be cooled to as low as −130°F without freezing. Below these temperatures, supercooled liquid freezes extremely rapidly into ice-Ih; this is called *homogeneous nucleation*. This phenomenon has frustrated experimenters wishing to investigate the properties of supercooled water, as well as amorphous ices, since all these phases can exist under similar conditions. Observing changes between these various phases would provide valuable information about water's anomalies, and help answer questions about the structure of liquid water. Since the region of the phase diagram in which all of these changes happen has proved unreachable for decades, it has been dubbed *no man's land*. Several studies have made small inroads into this uncharted territory, but most of them involved preventing the water from freezing by using extremely small amounts of it or by mixing it with other compounds—so any conclusions made might not apply to "bulk" pure water.

In 2014, however, a team of scientists using the Linac Coherent Light Source, the first X-ray laser in the world, managed a more meaningful exploration of no-man's land.[22] They produced a stream of droplets that cooled by evaporation, but remained large enough to be considered bulk liquid. The droplets cooled to temperatures just inside no-man's land, but remained liquid. The scientists were able to examine the structure of the water using X-ray pulses that were just 50 femtoseconds long (5×10^{-14} seconds). They found that water has more structure (a greater degree of tetrahedral bonding) the lower the temperature. The structuring reached a peak at the homogeneous nucleation temperature, which suggests that the ordered structure of water molecules provides its own nucleation surfaces—something that had previously been suggested by computer simulations.

One of the likely features of no man's land, that researchers would be keen to explore, is a transition between two different phases of liquid water—the hypothetical two structures that are both supposed to be present in liquid water at everyday temperatures. There may even be a critical point at the end of the line that marks that transition (like the critical point at high temperature and pressure for liquid/gas, above). Many researchers believe that the hypothetical two structures of liquid water (low- and high-density) may be directly related to the two main phases of amorphous ice (LDA and HDA).[23] In 2015, H. Eugene Stanley and

Phase diagram showing the relationship between ordinary liquid water, supercooled water, low- and high-density amorphous ice and the region known as no man's land. The green line extending into no man's land is the hypothesized *liquid–liquid phase transition*, which may delineate between low- and high-density forms of liquid water. At its end is a hypothesized *liquid–liquid critical point*.

Francesco Mallamace gathered together evidence from experiments and computer models to make the case for the existence of the phase transition between two different phases of liquid water (in no-man's land).[24] Further exploration of no-man's land, and the links between supercooled and amorphous water, may soon bring a definitive answer.[25]

While many mysteries remain, it is clear that liquid water possesses some kind of internal structure—it is heterogeneous. Its structure is effected by a dynamic network of hydrogen bonds. Those bonds are central to what makes water so much more than just two parts hydrogen and one part oxygen—and one can wonder if D. H. Lawrence was aware of them. Hydrogen bonds play a vital role in how water interacts with other substances—which is the subject of the next chapter.

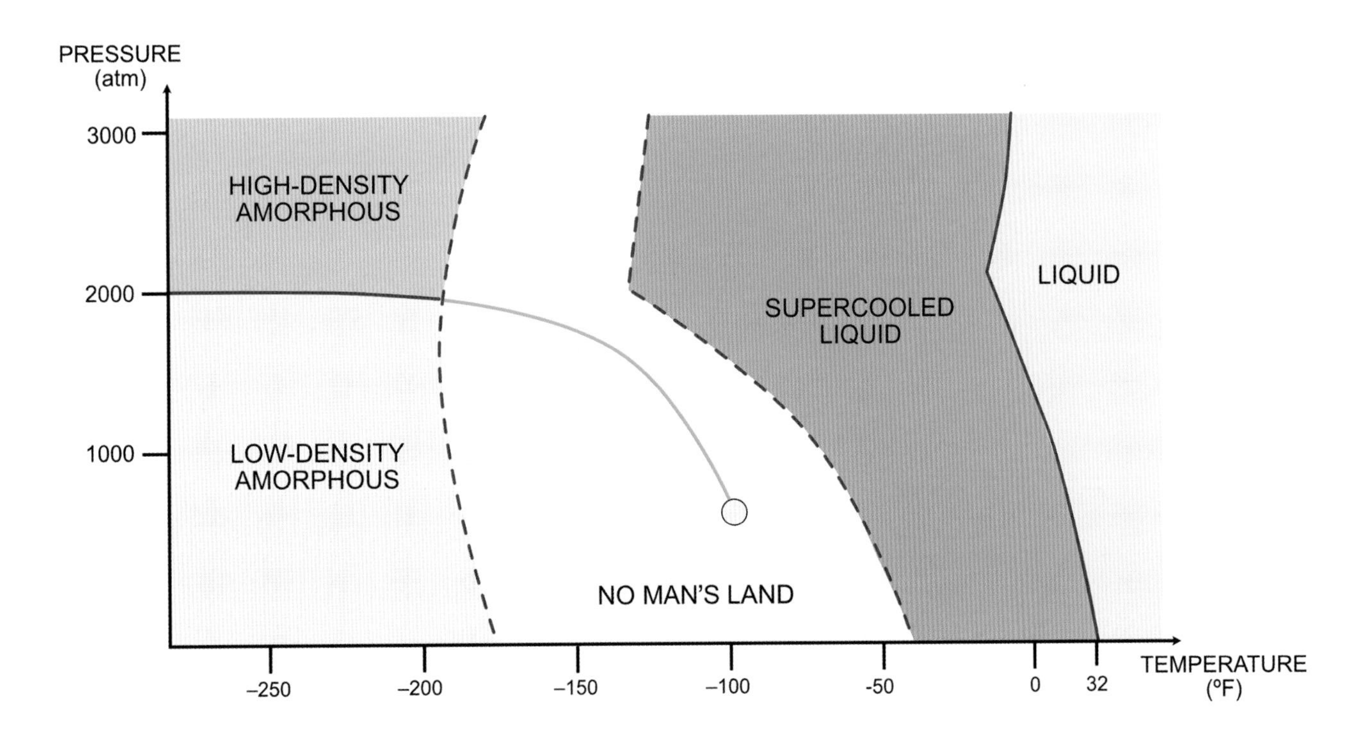

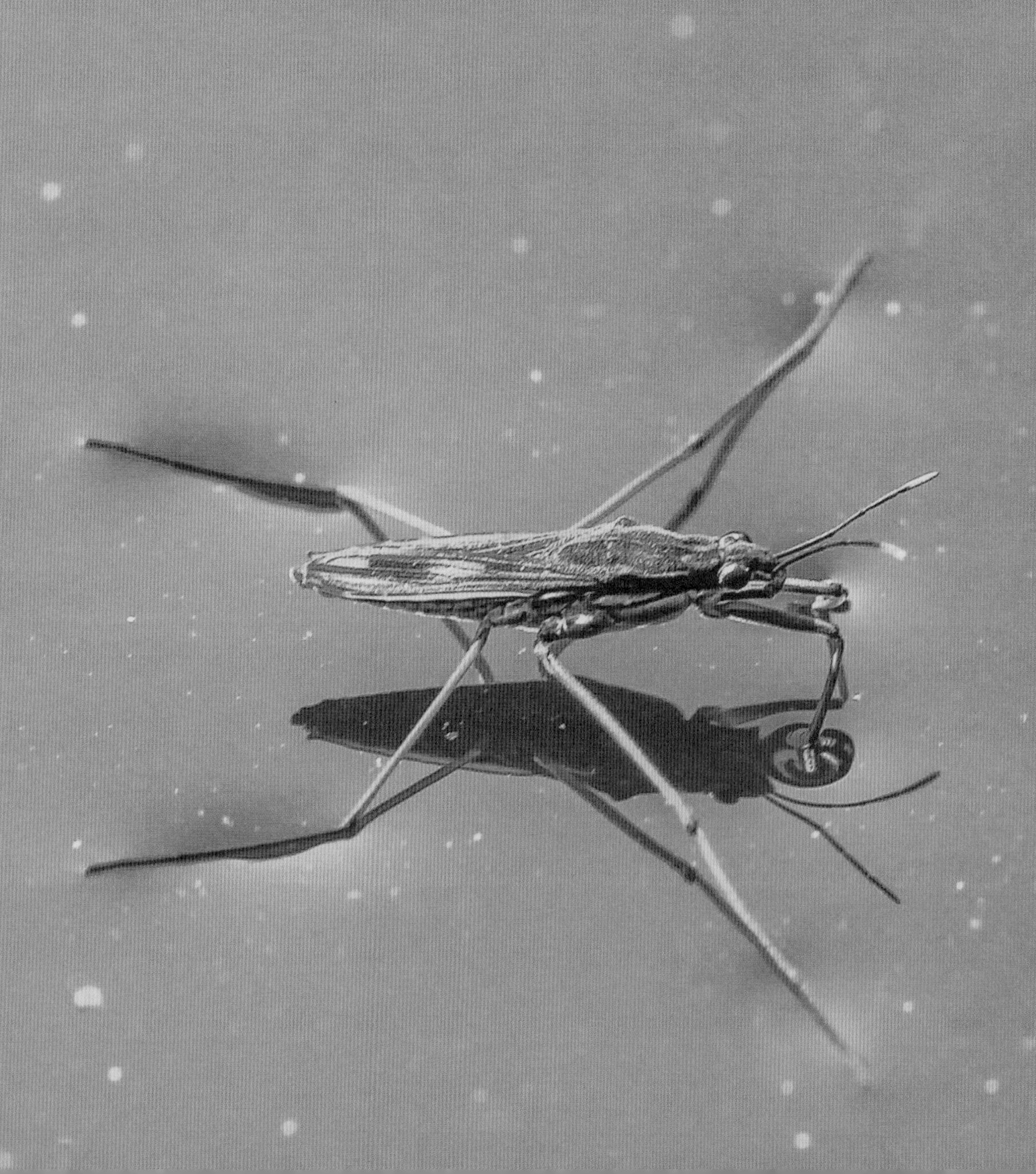

CHAPTER 5

At Water's Edge

"**Often borders** are thought of as passive objects, or matter-of-factly just as edges. However, a border exerts an active influence." Writer and activist Jane Jacobs wrote these words in her acclaimed 1961 book *The Death and Life of Great American Cities*. As for cities, so for water: it is at its interfaces with other substances that this ubiquitous and vital substance is most influential, most active. The surface of liquid water, for example, is where oxygen and carbon dioxide gases enter and leave—crucial to Earth's biosphere; it is the surface of ice that makes it slippery; it is at the interface between water and solid substances where water wets, or does not wet, where metals corrode. There can also be borders of a kind inside the bulk of liquid water—between water and substances that will or will not mix with it or dissolve in it. The fact that water is made of small, polar molecules (with an unequal distribution of electric charge) means that it dissolves a huge number of other elements and compounds—so much so that water is often called "the universal solvent."

Breaking the Surface

Water's surface is what gives water its feel—sticky but slippery, as the molecules adhere to fingers but also flow over each other. The surface of liquid water is also a place of frenetic interaction between water and air, as molecules of both leave and join the surface. With its propensity for dissolving and releasing gases, the interface between water and air is of great importance in biology, chemistry, and environmental and climate science. Since water's surface is exposed and accessible, there is ample opportunity to study it—and studying the precise nature of the water–air surface could help inform engineers, geologists and materials scientists on the behavior of the interface between water and other materials. Unfortunately for them, the structure of water's surface is every bit as tricky to work out as the structure inside liquid water, if not more so.

The tetrahedral bonding pattern of water molecules in solid and liquid water, explored in chapter 4, is disrupted, truncated, at the water–air or ice–air interface. At the surface there are molecules below and to the side, but none above, so the average number of hydrogen bonds per molecule is reduced. As a result, there are many *dangling bonds*—potential bonding sites that are unfulfilled (in this case, potential hydrogen bonds). Each dangling bond exposes either the partial negative charge of a lone pair of electrons on an oxygen atom or the partial positive charge of a hydrogen atom; dangling bonds make surface water more reactive than water in the bulk. Depending on their orientation, water molecules at the surface may have their hydrogen atoms or their lone pairs sticking up. Studying the dangling bonds gives the best insight into the character and structure of the surface of water.

When hydrogen bonds are absent, the O-H covalent bond becomes shorter and stronger. A technique called *sum-frequency generation spectroscopy* (SFGS) is sensitive to the O-H bond length, and can give scientists a good idea of the proportion of molecules that are hydrogen bonded, as well as their orientations. SFGS can also be used to investigate the librational mode[1] (see chapter 3) in which molecules twist back and forth around hydrogen bonds. It is the most important and effective tool for scientists investigating the nature of water's liquid–air surface. SFGS typically involves exposing the surface to a beam of infrared radiation that can be tuned across a range of frequencies. At certain frequencies, the incoming radiation will resonate with the stretching and relaxing of the O-H covalent bond in water molecules (one of the modes of vibration explored in chapter 3). At the same time, visible laser light of a fixed frequency shines onto the surface. The two beams interact, and the resulting sum of their frequencies is recorded, forming a spectrum that holds information about the bonds. This technique is particularly useful at surfaces, because it excludes contributions to the spectrum from molecules in the bulk liquid. Although SFGS is a practical experimental tool, used to investigate real water, it also offers insight into the surface structure when it is simulated in computer models. Such models can be used to construct a virtual SFG spectrum—and particular features of the spectrum can be traced back to molecular interactions or orientations within the model. Then, when those features appear in the real SFG spectra, experimenters can interpret them with confidence.[2]

Model of the very top surface of liquid water. Molecules at the very top, with unfulfilled, or dangling, bonds, are more likely to have one of their hydrogen atoms pointing up and out of the surface than their oxygen atom.

The results of SFGS and simulations of the surface of liquid water suggest that the average number of hydrogen bonds per molecule is very slightly above 3.[3] Inside liquid water, the average number of hydrogen bonds per molecule is closer to 4 (typically around 3.7). Molecules at the surface are far more likely to have the H of their O-H covalent bonds pointing up toward the air. The hydrogen bonds that do form at the surface create a largely two-dimensional network, with four-, five- and six-membered rings of water molecules[4]—so the surface of liquid water has some structure, some sense of order, despite the fact that water and other molecules are constantly leaving and joining the liquid there. Despite this apparent two-dimensional layer at the very top, the surface of liquid water is not a sharp border between the air above and the liquid bulk below. There exists a distinct layer, whose exact depth is difficult to determine and define, in which there is a decrease in density from the bulk liquid to the air and a variation in several properties of the water, over at least a nanometer (10 Å, equivalent to about three water molecules end-to-end). The very top of the surface carries a small positive electric charge, likely due to those hydrogen atoms at the end of the O-H dangling bonds—but overall, the surface layer carries a slight negative charge, while the bulk liquid is slightly positive. No one is quite sure of the origins of this charge separation, or its exact nature. The influence of those deviations from the properties of the bulk liquid extends many nanometers below the immediate surface. But much is still not known about the surface layer on

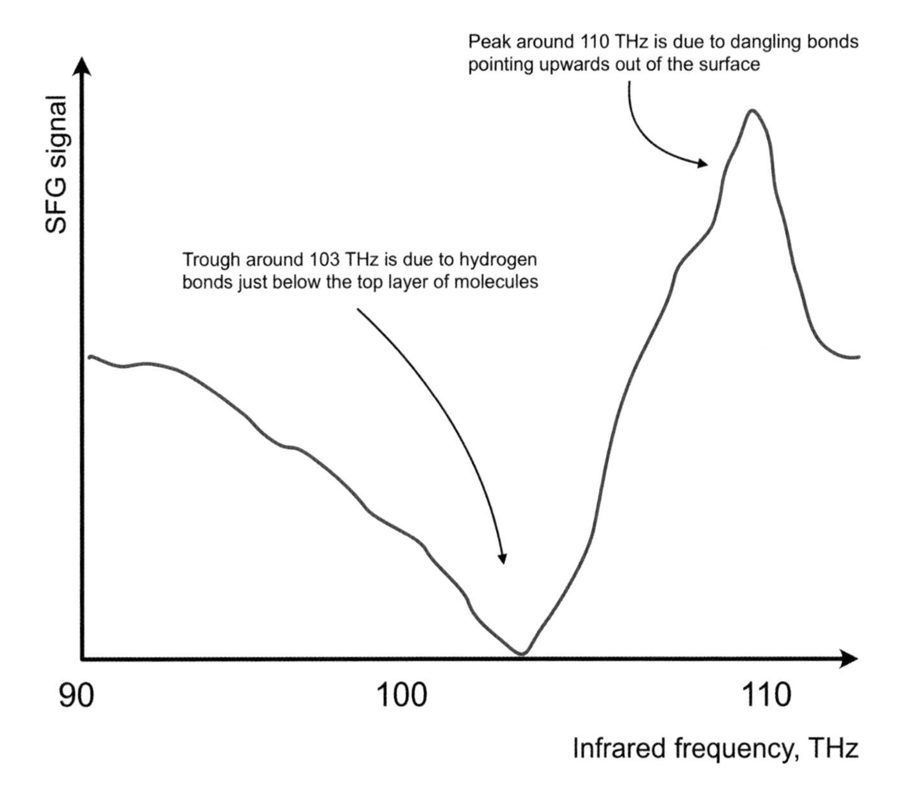

Mock-up of a simplified SFGS spectrum of the surface of liquid water. The horizontal axis represents the frequency of the probing infrared radiation in terahertz, or trillions of undulations per second. (These values are converted from a quantity that spectroscopists normally use, called wavenumber.) The vertical axis shows the SFGS signal—again, spectroscopists would use a different quantity, called susceptibility. The curve itself is simplified, too, showing only the major peak and trough corresponding to hydrogen bond presence or absence above the surface, respectively, and orientation. Finer details in the spectrum give tantalizing, unresolved clues to the nature of water's surface.

top of water, and studies are being carried out all the time, in a quest to characterize and understand this important molecular environment, which we encounter every day of our lives.

Slipping and Sliding

Just as for liquid water, the surface of ice also has no sharp boundary, and is also the subject of experiment and debate. Here, the surface layer is what makes ice slippery. As noted in chapter 4, the slipperiness of ice is often attributed to the fact that ice melts under pressure—and that standing on sharp-bladed ice skates exerts enough pressure to melt the ice. According to this hypothesis, first put forward in the 1880s, the pressure melting would result in a thin layer of liquid water that would lubricate the surface, allowing the skates to glide gracefully.[5] Clearly, light objects such as hockey pucks, that do not exert much pressure, also slip and slide easily on ice—a major problem with this hypothesis. A different idea was put forward in the 1930s. In this alternative account, it is friction, not pressure, that melts the ice—again, producing a thin layer of liquid water that lubricates the surface, allowing the skates to glide gracefully.[6] But ice is slippery from the get-go—and yet there is no way it can melt in anticipation of the movement of an object on its surface. One part of both those explanations is true: there is indeed a liquid layer—or rather *quasi-liquid* (liquid-like) layer—at the surface of ice. But it is not produced by the pressure of someone standing on the ice, nor by friction: it is always there—at least, down to temperatures as low as –30°F. Below this temperature, ice is no longer very slippery, and at –150°F, ice is as un-slippery as rubber. It was English scientist Michael Faraday who first suggested the existence of a liquid-like layer, in 1859, before either of the explanations given above had been proposed—although Faraday was concerned with a different phenomenon (how two blocks of ice stick together).

The fact that you cannot see the quasi-liquid layer at the surface of ice is not surprising: it is very shallow. It is 12 nanometers deep (about 0.5 millionths of an inch) at –11°F, increasing to 70 nanometers deep (about 3 millionths of an inch) just below freezing.[7] The thickening of this layer with increasing temperature (up to freezing point, of course) means that ice's *frictional coefficient* (slipperiness) goes down as the temperature goes up—the opposite of what happens with nearly all other solids. Since its thickness increases as the temperature approaches the melting point, the quasi-liquid layer at the surface of ice is sometimes referred to as a pre-melting layer. But the layer is not actually caused by melting, by warm air above: it is present even when the temperature of the air above the surface is well below freezing.

Just as mysteries remain about the structure of the water–air interface, the exact nature of the quasi-liquid pre-melting layer at the surface of ice is as yet unknown. Measurements have shown that it is far more *viscous* than bulk water, and is considerably less dense than liquid water or ice. Several researchers have used sum-frequency generation spectroscopy to study the hydrogen bonding, and the disruption of it, in ice's quasi-liquid layer. A 2017 study that utilized experimental and simulated SFG spectra found evidence that hydrogen bonds break first in the uppermost layer, then the next layer down, then the next, and so on, as the temperature climbs toward the melting point.[8] A computer model used in a 2018 study suggested that the populations of doubly- and triply-hydrogen-bonded molecules in the quasi-liquid layer vary with temperature.[9] In that model, as temperatures increased toward the melting point, the population of molecules with just two hydrogen bonds grew. Those molecules were more mobile within the (virtual) surface layer, and their mobility provided the lubricating effect.

One fascinating idea, put forward in 2014, was that the surface layers on water and ice are identical—both lower density than either water or ice, both viscous, both elastic (stretchy).[10] The researchers behind the study suggest that "Neither liquid skin

Model of the very top surface of liquid water. Molecules in the quasi-liquid layer have broken or distorted hydrogen bonds, and dangling bonds are again more likely to have one of their hydrogen atoms pointing up and out of the surface.

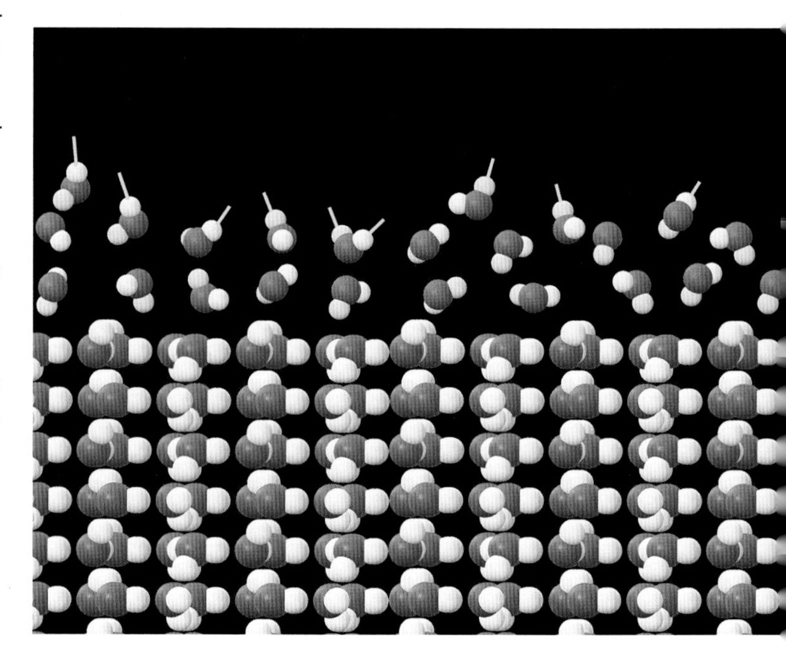

forms on ice nor ice skin covers water. Instead, a supersolid skin not only lubricates ice but also toughens the water skin." A *supersolid* is a material with a regular structure, like a solid, that can flow, like a liquid.

Pulling Together

Inside the bulk of liquid water, molecules can pulled in all directions by hydrogen bonds; the overall force, on the average molecule, is zero. But at the surface, molecules can be pulled only downward or to the side (they are not attracted upward, by the air). If you try to increase the surface's area, you need to bring more water molecules up from the bulk, which means fighting against those downward and lateral forces. This imbalance of forces gives rise to *surface tension*, an attraction that pulls any body of liquid water into the shape with the smallest surface area. Unrestricted by gravity or a container—for example, in an orbiting space station or in free-fall having just left a rain cloud—the shape with the smallest surface area is a sphere.

Surface tension is what makes a stream of water break up into small beads and droplets—a phenomenon called *Plateau–Rayleigh instability*. This is what happens to water from a garden hose, a shower head, a slow-running faucet or milk from a jug. Even if the flow begins as one steady, continuous stream, there will be very slight variations in its diameter. Where the diameter is slightly smaller, surface tension will begin to pinch off the stream, resulting in droplets. (A proper explanation of this phenomenon requires some detailed and rigorous mathematics, but this is the general idea.) Plateau–Rayleigh instability is put to good use in inkjet printers, where water-based inks break up in a controlled way after leaving a nozzle, *en route* to the paper.

Surface tension acts like an elastic skin on water—similar to the rubber of a party balloon. Its action can create small ripples, called *capillary waves*, whenever the surface is disturbed. When water's surface is pushed down—by a small object causing a splash, for example—the area of the surface is increased. Surface tension

acts to restore it, pulling it back level. The momentum of surface tension's restoring action carries the surface above its original level—just as when the momentum of a pendulum takes it past its vertical midpoint (gravity is the restoring force in that case). The surface tension acts to restore the surface once again, this time pulling it downward—and again, the surface plunges past level. The periodic, or regular, repeating, motion of the water's surface, up and down, affects the adjacent part of the surface, and the whole process repeats over and over, until its energy is dissipated, in small waves of ever decreasing height. Surface tension is the only relevant restoring force in capillary waves, which can have wavelengths up to about 11/16 inch. There are much tinier capillary waves at the surface of water, caused only by tiny molecular fluctuations—but they are enough to ensure that water's surface is rough at the nanoscale. As a result, the surface area is greater by about 15 percent than it would be if it were perfectly flat. Surface tension also plays a crucial role in bigger waves, called gravity waves—but in that case, as the name suggests, gravity is the dominant force that restores the surface to level (and beyond).

Astronaut Karen Nyberg, aboard the International Space Station, watching a glob of water floating free in front of her face. The water's surface tension pulls the drop into the shape with the smallest surface area, in this case a sphere. *Source:* NASA.

You have surface tension to thank for every sand castle you have ever built. The sand-to-water ratio is crucial: too little water and the castle will collapse, too much and the mixture is too fluid, like a soup. When the ratio is just right, the water forms tiny bridges between adjacent grains of sand—and surface tension, acting like an elastic band, pulls them together. This kind of liquid bridge explains why a glass tumbler with a wet bottom will lift a beer mat, and why the hairs on a wet head cling together.

You can also observe the effects of surface tension in a glass of wine or spirits. The thin streams of liquid that run down the inside of the glass after you swirl the

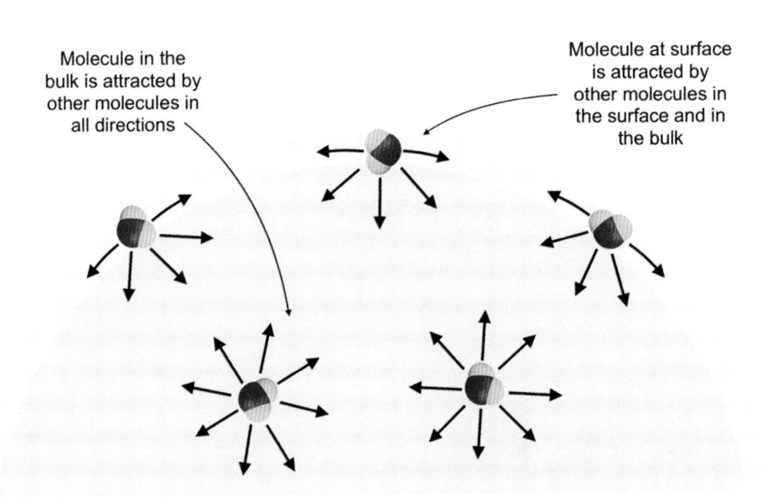

A simplified representation of the forces on a molecule inside the liquid versus one at the surface.

glass around are sometimes called "tears of wine." They are caused by something called the *Marangoni effect*—a phenomenon that happens whenever there is a gradient of surface tension, from high to low. An alcoholic drink is mostly a mixture of alcohol and water. Like any solution, its surface tension is less than that of pure water (more on that later in this chapter). Swirling creates a thin film of the mixture on the inside of the glass—and as alcohol evaporates much more rapidly than water, the concentration of alcohol there decreases compared with the mixture in the bulk of the liquid wine. The increased surface tension in the thin film drags more of the mixture up from the bulk—the wine climbs up the glass. Eventually, the amount of liquid raised above the level surface of the wine is so great that it becomes unstable; gravity wins out, and begins to pull it back down—and, of course, surface tension causes it to contract into rivulets and droplets.

Two effects of surface tension can be seen in this photograph of a water drop causing a splash: the recoil creates a rising body of water that breaks up into droplets, thanks to Plateau–Rayleigh instability, and capillary waves (ripples) and gravity waves travel outward from the site of the disturbance. *Source:* Milk Sculpture. Credit: Stefan Eberhard.

Sticking to the Sides

There is another important force at work in sand castles and in tears of wine, besides surface tension (and besides gravity, of course). Just as water *coheres* (is attracted to itself), so it also *adheres* (is attracted to certain other materials). If water did not adhere to sand grains, then those liquid bridges would not form, and the sand and water would remain completely separate. If the wine did not adhere to the side of the glass, it would fall straight back down, giving no opportunity for the tears to form. The cause of adhesion is very similar to that of cohesion: electrostatic attraction. Water adheres most strongly to surfaces that are polar—where there is a variable distribution of electric charge, as explored in chapter 3. Both sand and glass are mostly silicon dioxide, and the bonds between the silicon and oxygen atoms are polarized: again, the negative electric charge is more

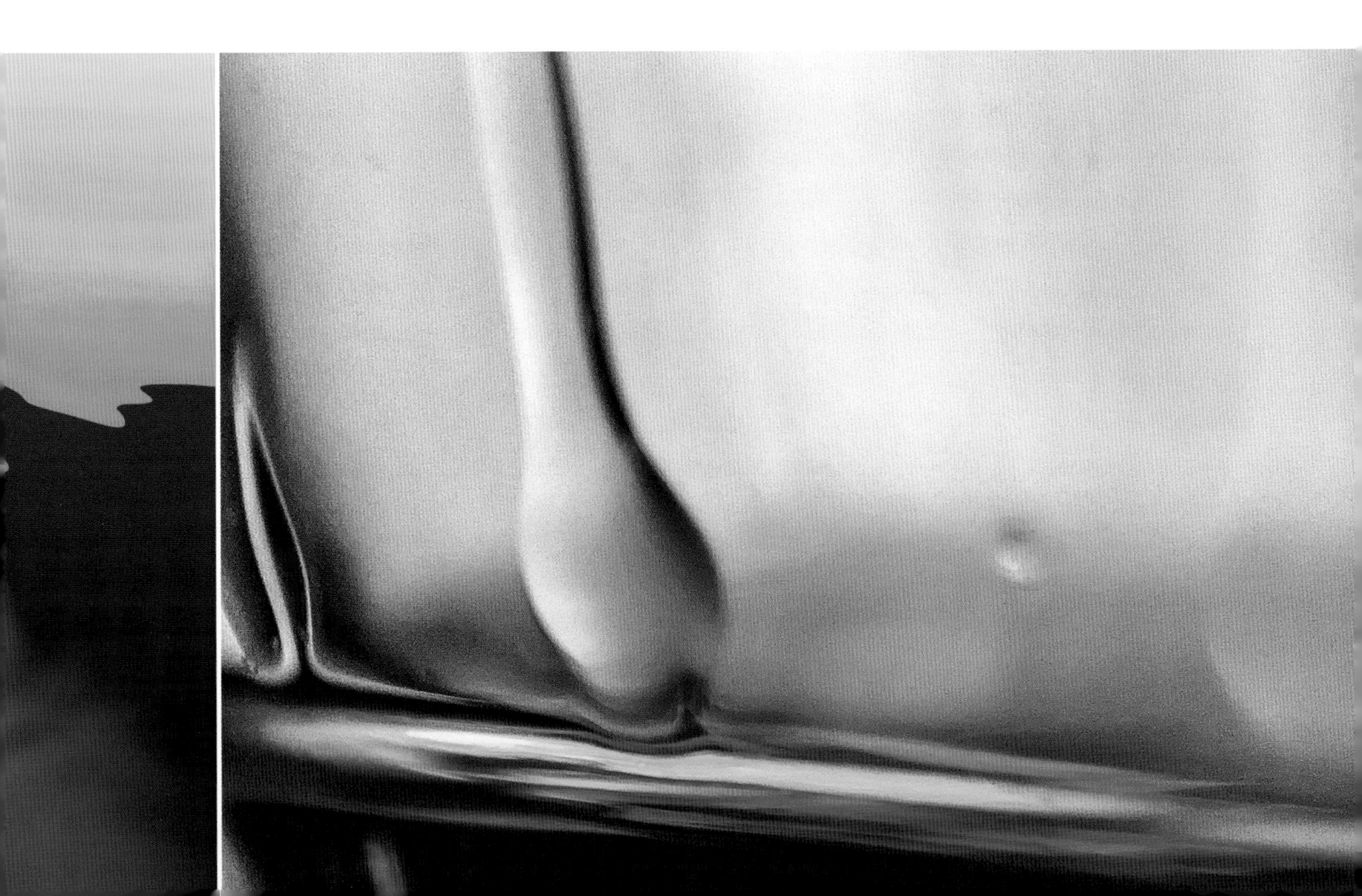

Tears of wine—in a glass of blue Curaçao. As alcohol evaporates from the thin layer of liqueur left on the side of a wine glass, the increased surface tension in the less-concentrated solution that results drags more liquid up the glass, until gravity pulls it back down again. Surface tension pulls the descending liquid into rivulets and droplets.

concentrated around the oxygen atoms. So, at the surface of glass or of a grain of sand, the partial positive charge at the hydrogen atoms in a water molecule is attracted to the partial negative charge of the oxygen atoms in the silicon dioxide. This is why water curves up where it meets the edge of a glass of water—the resulting curve is called a *meniscus*.

The combination of adhesion, cohesion and surface tension is responsible for another important phenomenon: *capillary action*. Place the end of an open tube into the surface of water, and the water will climb up the side of the tube, pulled by adhesion with the glass. Meanwhile, the surface tension pulls the water's surface taut, balancing the force of gravity that is pulling the column of liquid downward—and the bulk of the column of water stays together thanks to cohesion. The narrower the tube, the higher the column of liquid can climb. Capillary action not only occurs vertically. For example, tears, which keep the eye moist, drain by capillary action through two very thin tubes (canaliculi) into a small chamber (the lacrimal sac), from where they empty

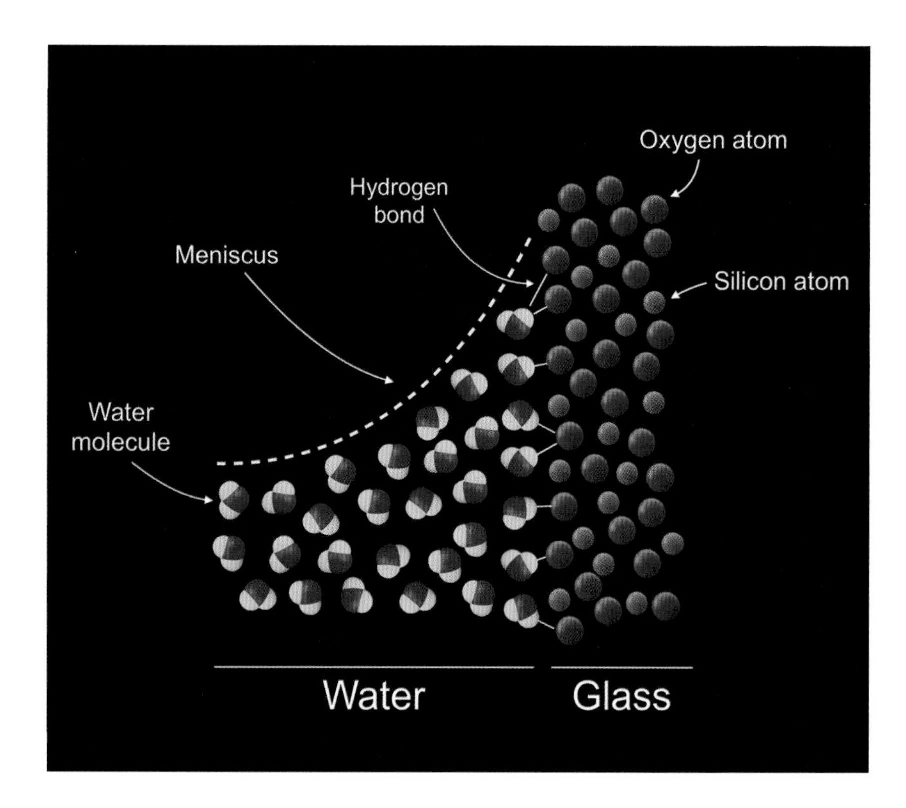

Molecular scale illustration of water adhering to glass. The oxygen atoms in the glass carry a partial charge, which attract the partial positive charges of the hydrogen atoms in nearby water molecules.

▲ **Capillary action** in action. The thinner the tube, the higher the water climbs, pulled up against gravity by adhesion between water and glass.

into the nasal cavity (of course, when you produce more tears than this system can deal with, the water will run down your cheeks). Capillary action is also responsible for the wicking of water (or water-based solutions, such as spilled wine) into paper or cotton towels. Paper and cotton are composed of hollow fibers of cellulose, which act as extremely thin capillaries. Those same thin, hollow fibers help lift water from roots to leaves in plants—as explored in chapter 6. And capillary action is also responsible for drawing paint up into the bristles of fine brushes.

Water Hating

Water does not adhere to all materials as well as it does to glass and to sand. The capillary action experiment described above, with an open tube poking out of water, can turn out very differently if you use a tube made of something other than glass. With a tube made from *hydrophobic* material (from the Greek for "water fearing"), the meniscus in contact with the tube can bend downward, rather than upward—and the top of the water column can retract beneath the surface of the bulk liquid, rather than being pulled up above it.

A drop of water placed onto a surface made of a hydrophobic material will not spread out. The interior angle the drop of water makes with the surface is called the *contact angle*, and a hydrophobic surface is one for which the contact angle is greater than 90°. That is why water droplets bead up and stand tall on a waxed surface, for example. The term *hydrophobic* is a bit misleading: a surface does not actively repel the water. Instead, water is simply attracted to itself far more than it is to the surface. Hydrophobic surfaces are not *wettable*, for this reason. If the water drop's contact angle is less than 90°, the surface is *hydrophilic* ("water loving"). Put a drop of water on a hydrophilic surface and it will spread out a fair bit, forming an extensive, shallow pool. At zero contact angle, water completely wets a surface.

A leaf of the lotus plant (*Nelumbo nucifera*), with water not wetting the surface, thanks to the leaf's superhydrophobicity—a phenomenon sometimes referred to as the lotus effect.

Some surfaces are extremely hydrophobic: if the contact angle a water drop makes is greater than 150°, the surface on which it rests is *superhydrophobic*. Certain insect wings and plant leaves are superhydrophobic, thanks to a hydrophobic material at the surface layer and a complex structure made up of millions of tiny protrusions, which ensure a water drop makes minimal contact with the surface. Water runs off such a leaf or wing extremely easily, taking with it potentially harmful bacteria and fungi—and, in the case of the leaf, dirt that would obscure the leaf from sunlight it needs for photosynthesis. Some materials engineers design surfaces that mimic this self-cleaning ability. Coatings are currently available for glass that wash away dirt and prevent water films, which would otherwise deposit dissolved minerals when they evaporated. Solar panels coated with such superhydrophobic materials require much less maintenance.

Aquatic insects make use of superhydrophobicity, too. Water striders (family *Gerridae*), for example, have very hydrophobic legs, which allow them to "walk on water": they dent but do not break the surface, and the restoring force of surface tension is enough to support the insect's weight. The legs of a water strider are covered with tiny rigid hairs, each one decorated with minuscule indentations—all of which makes them superhydrophobic, helping repel water.[11]

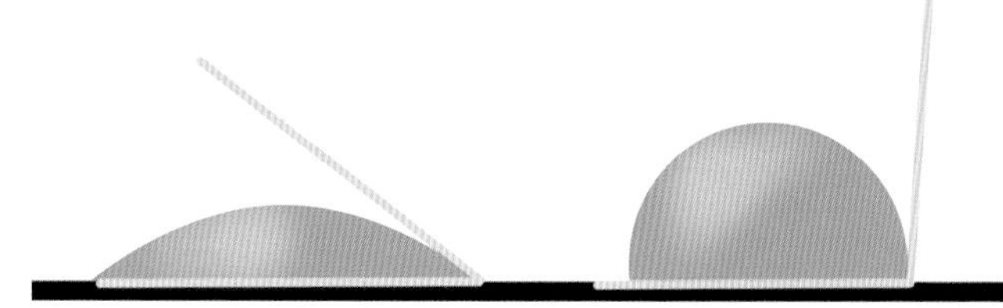

Hydrophilic
<90º
Wetting
Hydrophobic
>90º
Non-wetting
Superhydrophobic
>150º
Non-wetting

Underwater Mysteries

Just as there are unanswered questions about the nature of the water–air and ice–air interfaces, the boundary between a submerged solid surface is also the subject of mystery, experiment and debate. When the boundary surface is hydrophobic, studies reveal that immediately adjacent to the surface—for the first 5 Å (20 billionths of an inch)—water has a density less than half that of ice, and an abundance of free, dangling, non-hydrogen-bonded O-H covalent bonds.[12] Beyond that first layer is another, about seven times as wide, in which the degree of hydrogen bonding is closer to that in bulk liquid water, but with more order and, again, lower density. These differences in the properties of water at the hydrophobic boundary and water in the bulk are due to the truncation of hydrogen bonds at the surface, in a similar way to the air–water surface described above. In fact, the air at the water–air surface can be considered as a hydrophobic surface, as water is not attracted to it, and each situation can inform the other.

The relationship between contact angle and hydrophilicity, hydrophobicity and superhydrophobicity.

It is at the boundary between water and hydrophilic surfaces that things become really interesting—and puzzling and controversial. Here, the network of hydrogen bonds is nowhere near as truncated, since water molecules are attracted to such surfaces, via hydrogen bonds. And yet a boundary layer forms that has very different properties from the bulk liquid—and this layer extends much further than in the case of the submerged hydrophobic surface—much, much further. In 2003, American professor of bioengineering Gerald Pollock suspended tiny latex spheres in water, and submerged a surface of Nafion—a hydrophilic polymer developed by DuPont in the 1960s. Astonishingly, within less than a minute, the latex spheres migrated away from the surface, leaving a clear layer, from which they were excluded.[13] This *exclusion zone* layer can extend several thousandths of an inch—more than 10,000 times the width of the layer next to a hydrophobic boundary. The exclusion zone can grow wide enough to be visible to the naked eye. Pollack found that the water in this exclusion zone, which he called *EZ water*, has strikingly different properties from bulk liquid water. Its refractive index, or how much it bends light,

A water strider on a pond. Surface tension pulls upward where the water's surface is indented, and the insect's hydrophobic feet prevent it from breaking through the water.

is significantly higher; it carries a negative charge throughout; it is more dense and more viscous. So distinct were the properties of EZ water that Pollack declared it to be a separate phase of water. Other researchers repeated Pollack's procedure, with the same result: the exclusion zone is a genuine phenomenon. In fact, experimenters have been aware of its existence, adjacent to a variety of hydrophilic surfaces, for decades. For example, in 1970, researchers reported an exclusion zone next to rabbit corneas 15 thousandths of an inch deep.[14] In 1978, researchers reported an exclusion zone around a large, membrane-bound organelle within the cell, called the Golgi apparatus—from which other, smaller, organelles, as well as large molecules, were excluded.[15]

The fact that the properties of exclusion zone water are so different from those of the bulk liquid led Pollack (and some before him) to suggest that the water is highly ordered—more structured than the liquid. Specifically, as far back as 2003, Pollack suggested that EZ water is formed by layers of water molecules bound together hexagonally in sheets, those sheets forming one at a time, pushing the latex spheres away. To account for the width of the exclusion zone layer would necessitate around a million such layers. In 2013, Pollack wrote a book,[16] setting out his observations and his suggested molecular structure for EZ water. He also suggested that, since there is a charge separation between (negatively charged) EZ water and (positively charged) bulk water, one might be able to produce a current by connecting a circuit across the two—like terminals of a battery. He further suggested that EZ water might be the reason for the stability of the water bridge in the water thread experiment (see chapter 3). And he put forward the idea that the exclusion zone is created or enhanced by light falling on the surface, so that EZ water might one day form the basis of solar cells. Pollack went further, claiming that EZ water plays a crucial role in blood flow in mammals. He also proposed that EZ water brings health-giving qualities to spring water and glacial meltwater—and that the negative charge it carries is an essential part of keeping us healthy.

EZ water is certainly a fascinating topic, and its structure, if any, remains a mystery. But many of Pollack's claims of the behavior, health-giving qualities and future prospects of EZ water have been met with skepticism in the scientific community. A 2020 review of EZ water questions many of Pollack's conclusions, particularly those related to the structure of EZ water and its purported health benefits.[17] Several researchers have shown conclusively that Pollack's suggested layered structure for EZ water is impossible. Furthermore, experiments into EZ water almost always involve Nafion, which is known to absorb water and produce a kind of gel—this might well influence the behavior of water at the interface. Glass is hydrophilic, but there is no exclusion zone at the glass–water interface.

One reason the mainstream scientific community is wary of Pollack's self-proclaimed "departure from the mainstream science route"[18] is their collective memory of something that happened just a few decades back: the polywater controversy, "one specter that haunts water research today more than any other other."[19] It began in the early 1960s, when Russian physicist Nikolai Fedyakin experimented with water that had been confined in tiny quartz tubes, and found that the water changed to a very dense, extremely viscous form, with a greatly elevated boiling point. Another Russian physicist, Boris Derjaguin, was intrigued, and developed and improved Fedyakin's experiments. In 1966, Derjaguin reported this new form of water to scientists in the West, but few took much notice at first. It was a scientific paper published in 1969 that coined the name *polywater* that seems to have caught the imagination of scientists, the media and governments alike. So entranced were scientists and the media that, in 1970, the *Wall Street Journal* proclaimed that American scientists had "closed the polywater gap" between the Soviet Union and the United States. Hundreds of serious scientific papers were published in the late 1960s and early 1970s suggesting applications for, and possible structures of, polywater—some of the proposed structures were almost identical to the one Pollack suggests for EZ water, with sheets of hexagonally bonded molecules. In the end, it turned out that polywater was a gel created by the quartz dissolving into the water in the confined spaces.

Seeking Solvation

Quartz does not readily dissolve in water, which is why scientists at first ruled out the possibility of polywater being a mixture of pure water and quartz. Quartz is fairly unusual in its low solubility: water can dissolve more elements and compounds than any other substance. It is water's strong hydrogen bonding, together with the small size of its molecules, that makes it so adept at dissolving things.

Unsurprisingly, perhaps, *ionic* compounds—composed of positive and negative ions—dissolve best. An ionic solid is a crystal in which the oppositely charged ions are held in place by their mutual attraction. Table salt, sodium chloride (NaCl), is the archetype of ionic compounds. At the atomic scale, a grain of salt has a cubic structure: sodium ions (Na^+) are attracted equally by six chloride ions (Cl^-), which sit at the corners of a cube around it. (And each chloride ion is similarly attracted by six sodium ions, at the corners of a cube.) The attraction between the ions is strong: salt has a melting point of over 1,400°F—only above that temperature do the ions have enough kinetic energy to break free from each other. And yet water easily breaks down the structure, surrounding the individual ions and carrying them away into the bulk liquid—even when it is cold. When they surround a sodium ion, the water molecules align so that the partial negative charge of the lone pairs on the oxygen atom points toward the positive charge of the ion. The water molecules around a negatively charged chloride ion have the partial positive charge of the hydrogen atoms pointing inward instead. By surrounding ions in this way, water molecules shield the ions' electric charge, which ensures the ions will not rejoin with each other, and will instead remain in solution.

Solvent molecules surrounding ions in solution form *solvation shells*; when water is the solvent, these are called *hydration shells*. In some cases, the influence of a dissolved ion can extend beyond just one layer. The water molecules in the innermost layer are more attracted to the dissolved ions than they are to other water molecules. In the hydration shell of a sodium ion, for example, the innermost layer is typically populated by between six

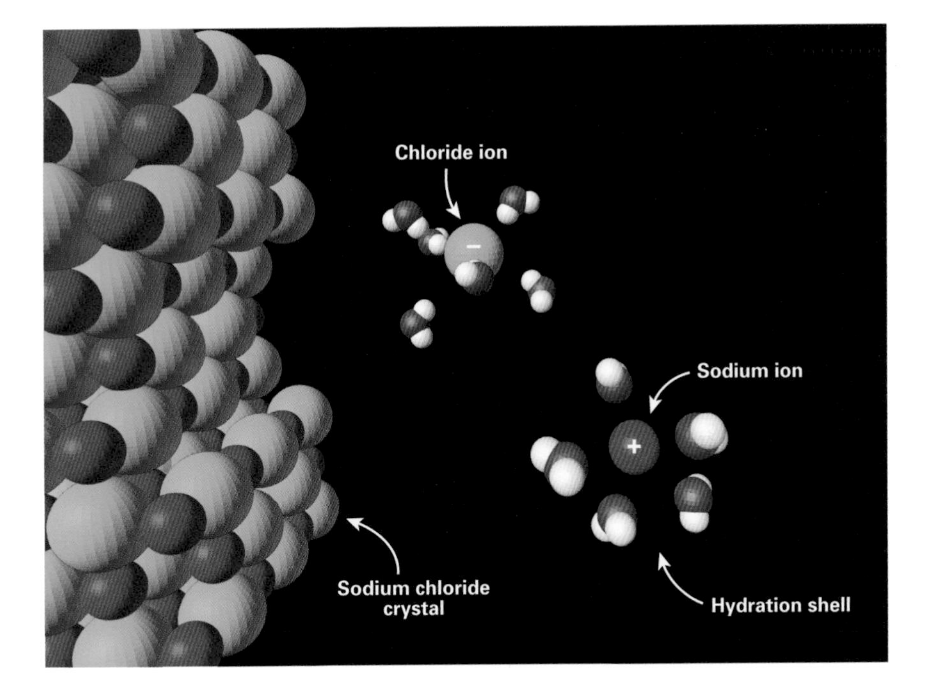

In a crystal of salt, sodium (shown violet) and chloride (green) ions are held in place by their mutual electrostatic attraction. When submerged in water, the crystal quickly dissolves, as water molecules dislodge the ions, surround them and carry them away. Notice how the partial charges on the water molecules are aligned with the opposite charge on the ion.

and eight water molecules. Of course, since water is a dynamic environment, molecules constantly leave and join the hydration shell—just as water molecules in clusters in liquid water swap in and out (see chapter 4). A 2016 study revealed how the presence of an ion in solution can affect the network of hydrogen bonds in liquid water far beyond the hydration shell, with a single ion influencing around a million water molecules.[20]

Water also dissolves itself, in a way. As explained in chapter 3, a small proportion of the molecules in pure water dissociate into hydroxide ions (OH^-) and hydrogen ions (H^+). The hydrogen ion cannot exist in solution: it is a naked proton with such a high concentration of charge that it sticks firmly to a water molecule, via hydrogen bonding, to form a hydronium ion (H_3O^+). The hydroxide and hydronium ions, themselves fractured parts of water molecules, are in solution among their unfractured siblings—and are therefore surrounded by a hydration shell. As they float around in the liquid, these dissociated, hydrated ions can take part in reactions, just as any ions can. For example, hydronium

The violent reaction between pure sodium metal and water. The sodium atoms give up their single outermost electron to hydronium ions in the water, leading to the evolution of hydrogen gas, which the heat of the reaction can ignite, and leaves behind an alkaline solution.

ions (H_3O^+) will easily accept an electron from a metal atom. Put sodium metal (Na) into water, and its atoms readily lose electrons, to become positively charged ions (Na^+). The electrons they lose are accepted by the hydronium ion, allowing the proton to escape and unite with the electron to become an atom of hydrogen (H). Two hydrogen atoms readily bond together, and the hydrogen molecules (H_2) that are produced join with others, to become hydrogen gas. The result is a solution that contains sodium ions and the hydroxide ions left over from the initial dissociation, but fewer hydronium ions (since hydrogen has been lost). If the concentration of hydrogen ions decreases from 10^{-7} to, say, 10^{-9}, the pH changes from 7 to 9, and this is an alkaline solution. Potassium, lithium, and cesium join sodium in group 1 of the periodic table, and all react violently with water, in the same way as sodium. As a result, these metals are called alkali metals, because of the alkalinity of the solutions they form when reacting with water.

Only the alkali metals react rapidly with water in this way. For other metals, the reaction is the same, but happens much more slowly. In an acidic solution, which contains many more hydronium ions (its pH is less than 7), the reaction can proceed more quickly. It was this reaction, with zinc and hydrochloric acid, that Henry Cavendish used most often to produce inflammable air (or hydrogen—see chapter 3).

It is not only ionic compounds and metals that dissolve well in water. Compounds made of molecules whose atoms are covalent but polar (as water is) also dissolve. Table sugar and alcohol are good examples. Both these molecules contain O-H covalent bonds—and, just as in the water molecule, there is partial negative charge around the oxygen atom at one end of the bond, and partial positive charge around the hydrogen atom. Water molecules form hydration shells around polar covalent molecules, as they do around ions.

Although water holds on to sodium, chloride and other ions, there is only so much it can hold. When molecules evaporate from seawater or salty lake water, the dissolved salts come out of solution. Here, workers at the Hon Khoi Salt Fields in Vietnam dump salt they have collected after it came out of solution in shallow evaporation ponds.

Oil and Water

Molecules that are nonpolar—which do not have areas of partial positive and negative charge—do not mix well with water. Oily and waxy substances are the best known examples. Drop some oil onto water, and the two liquids remain stubbornly separate. Vigorously shake a mixture of oil and water, and the oil will break up into small droplets. Let the mixture stand, and the oil droplets will slowly come together to form larger droplets, as hydrogen bonds bring water molecules closer together, nudging the oil molecules out of the way in the process.

That is not to say that compounds made of hydrophobic molecules can never mix well with water. Small hydrophobic molecules can fit into spaces within the structure of liquid water, the surrounding water molecules building cages around them as they struggle to retain their hydrogen bond *coordination*. These cages are open, like the structure of ice—that is how methane (a compound made of small hydrophobic molecules) forms the clathrate structures in methane hydrates, discussed in chapter 4. Carbon dioxide dissolves well in water, even though its molecules are nonpolar. There is a partial negative charge on the two oxygen atoms, and a partial positive charge on the carbon atom at the center, but the molecule is linear, not bent like water, so the polar effect cancels out. However, there is enough of a charge distribution to make carbon dioxide interact with surrounding, polar, water molecules. The carbon atom forms a weak, and long, hydrogen bond with a lone pair on an oxygen atom of a nearby water molecule, and the two oxygen atoms form weak, long, hydrogen bonds with hydrogen atoms on nearby water molecules. The interactions enhance the local tetrahedral bonding in the surrounding water, and create a cylindrical cavity in the water, with the carbon dioxide molecule at the center.[21]

There is another way in which hydrophobic molecules, such as oils, can be made to mix evenly with water. It involves compounds whose molecules can form a bridge between water and oily substances: they have one polar, hydrophilic end, and one nonpolar, hydrophobic end. Such compounds include sodium stearate, the most common compound in soaps. It dissolves in water to form (positively charged) sodium ions and negatively charged stearate ions. These stearate ions have a long tail of carbon and hydrogen atoms, which is hydrophobic, and a hydrophilic end, with oxygen atoms and a negative charge. Water molecules are attracted to the hydrophilic end—so the ion becomes hydrated—and fats and other hydrophobic molecules are attracted to the hydrophobic tail. A solution containing stearate ions (soapy water) can dissolve oils and other hydrophobic substances but still remain dissolved in water—thus carrying grease and dirt away when rinsed. Ions or molecules with hydrophobic and hydrophilic parts, including stearate ions, are described as *amphiphilic*. Another amphiphilic molecule is lecithin, which makes up nearly one-tenth of the weight of the yolk of a chicken's egg. This is why egg yolk can unite an otherwise frustratingly *immiscible* (unmixable) combination of (hydrophobic) olive oil and (water-based) vinegar, to make mayonnaise.

Amphiphilic compounds tend to aggregate at the air–water interface, with their hydrophobic tails sticking out of the surface (away from the water in the bulk liquid). This has the effect of reducing the water's surface tension, so the compounds are called *surfactants*. A soap solution has low enough surface tension that it can be drawn out into a thin film. A soap film is a very thin sandwich: the stearate ions, with their hydrophobic tails sticking out of the film's surface, are the pieces of bread, and water is the filling. When air is trapped inside these soap films, they make bubbles.

Stearate ion, with its long hydrophobic tail of carbon (black) and hydrogen (white) atoms and its hydrophilic end, made of an oxygen atom and an oxygen ion (both red) . Two representative water molecules are also shown, with their (partially positive) hydrogen atoms attracted to the negative charge of the hydrophilic end.

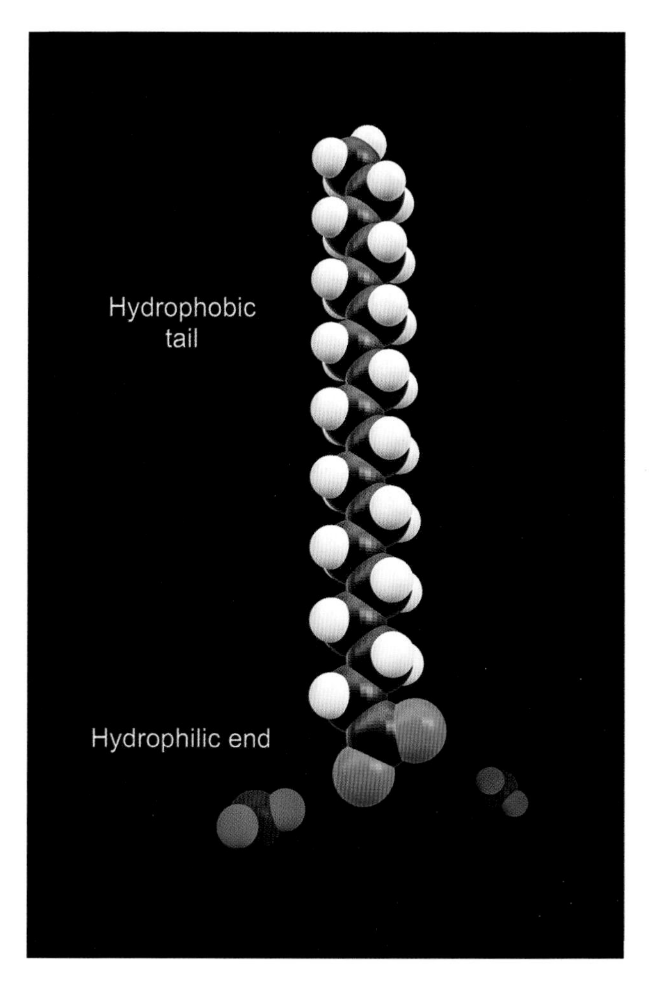

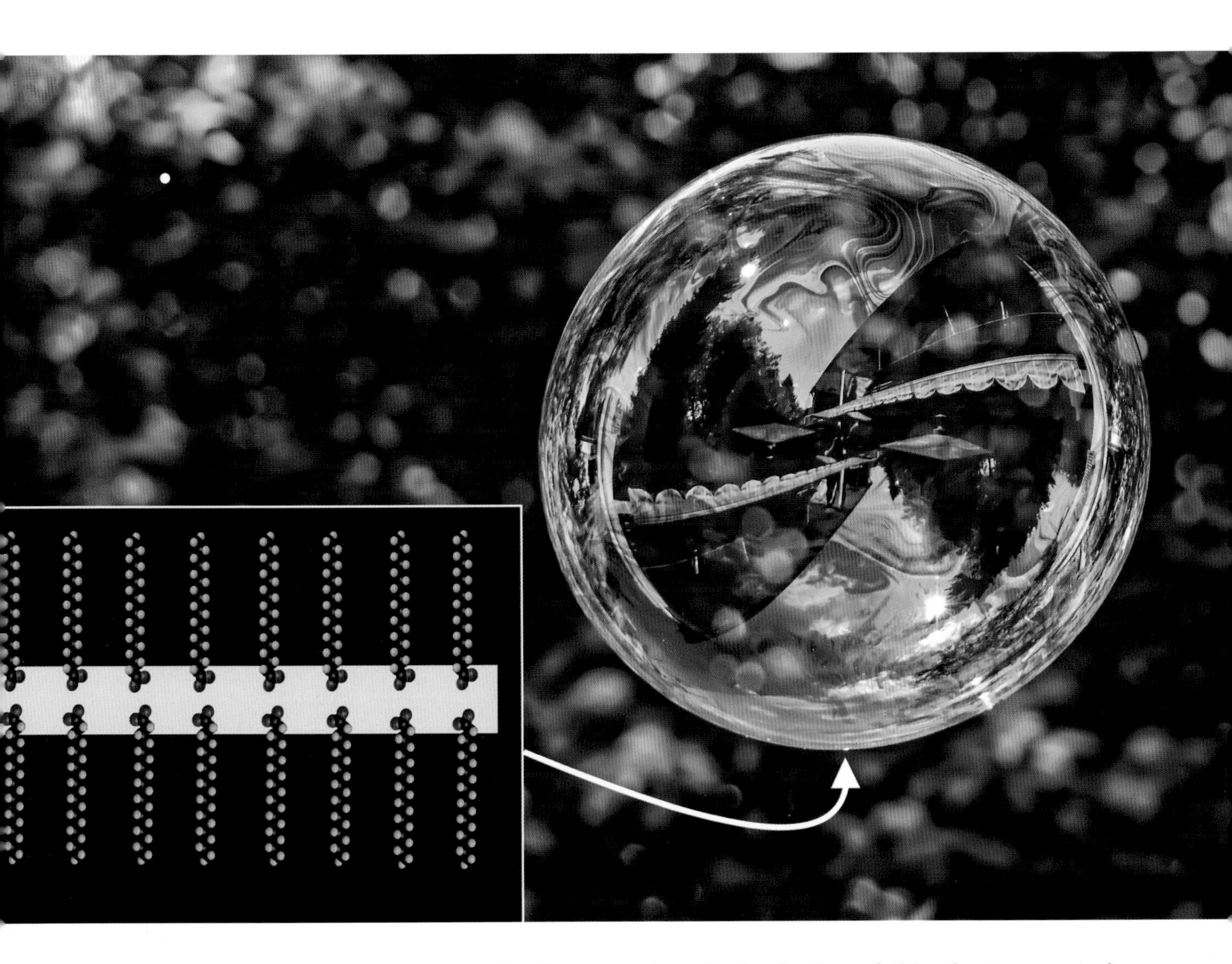

A bubble, made of a soap film surrounding air. Without surfactants, the water's surface tension would very quickly pull this bubble into a drop, or a few drops. The width of the soap film is typically around 50 and 500 nanometers (2 to 20 millionths of an inch)—so the inset schematic diagram, which shows how the stearate ions sit at the water's surface, is not to scale.

Jane Jacobs, quoted at the beginning of this chapter, wrote in 2004: "To science, not even the bark of a tree or a drop of pond water is dull or a handful of dirt banal. They all arouse awe and wonder." In each case, water is crucial in making these everyday things wonderful—and to enable us to wonder at them. For water is not just a bit player in the living world: the properties of water explained thus far are what make life possible on our bountiful, blue planet.

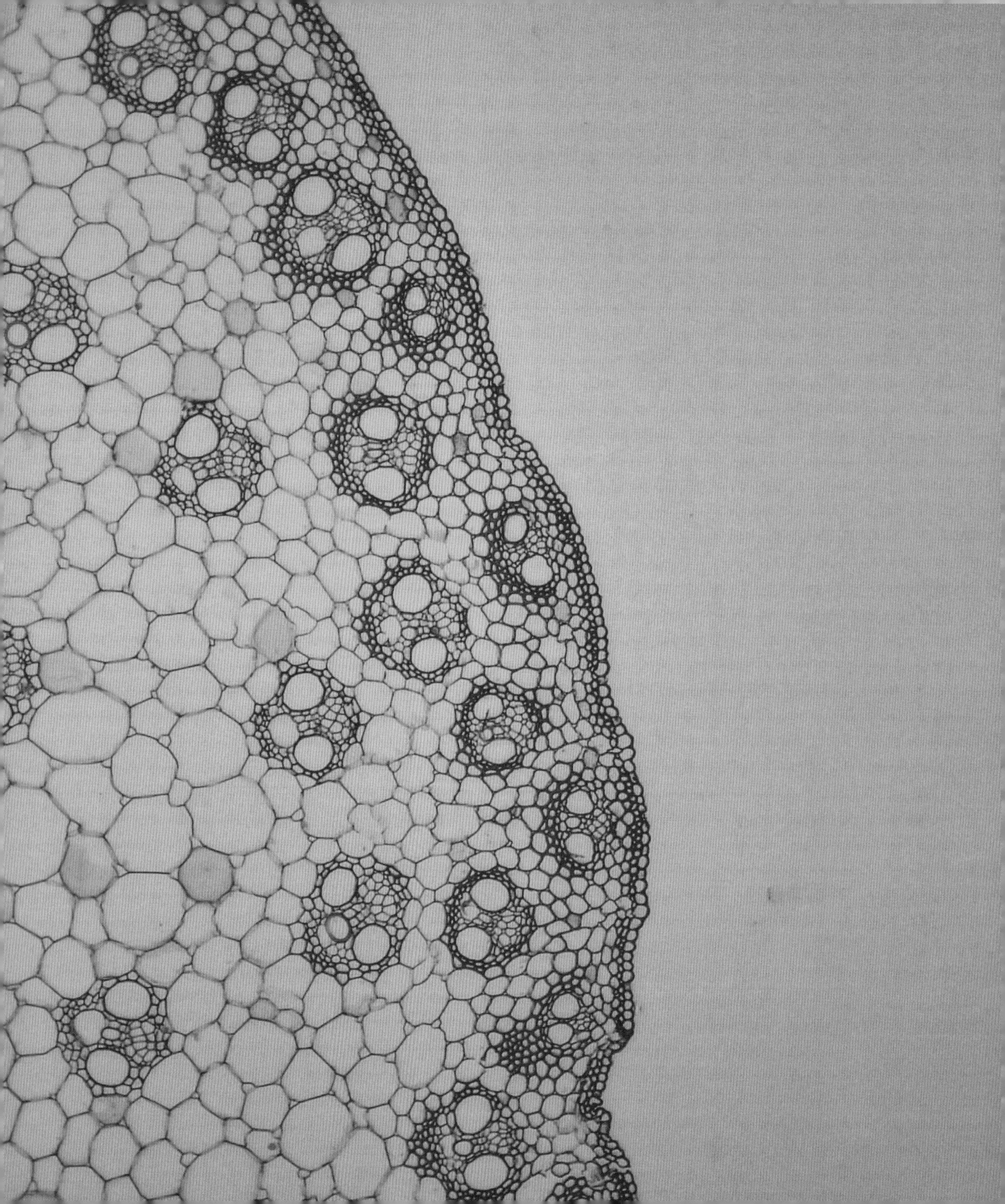

CHAPTER 6

"The Hub of Life"

In his 1972 book "The Living State," the Hungarian biochemist Albert Szent-Györgyi wrote, "Sixty years of research has taught me to look upon water as the hub of life." It is common to think of water as just the canvas for the biochemical processes that make living things tick—as a passive medium whose only role is to carry and mix the chemical compounds whose complex interactions comprise life. Water's excellent ability to dissolve substances is, of course, vital. But water does so much more: it provides the driving force for the formation of cell membranes, helps fold protein molecules and is intimately involved in the processes by which living things store and access energy, and in many of the reactions that build and break down the large molecules that give living organisms their physical structure. Without water's unique set of characteristics, there would be no life on our planet.

Life in the Bag

Every living thing on Earth—and therefore every living thing we have ever encountered—is comprised of one or more cells. (Viruses are not made of cells, but then they are not alive.) In the case of a single-celled organism, the cell is the entire organism. In a plant, animal or fungus, made of many cells, organs are made of tissues, which are comprised of different types of cell—so cells are the building blocks of these multicellular organisms. A cell is like a tiny closed bag filled with chemical compounds dissolved or suspended in water. All of the complex chemical reactions that constitute life take place in this enclosed, aqueous environment: for example, DNA is copied, and proteins are manufactured; in plant cells, light energy is captured and stored in carbohydrates;

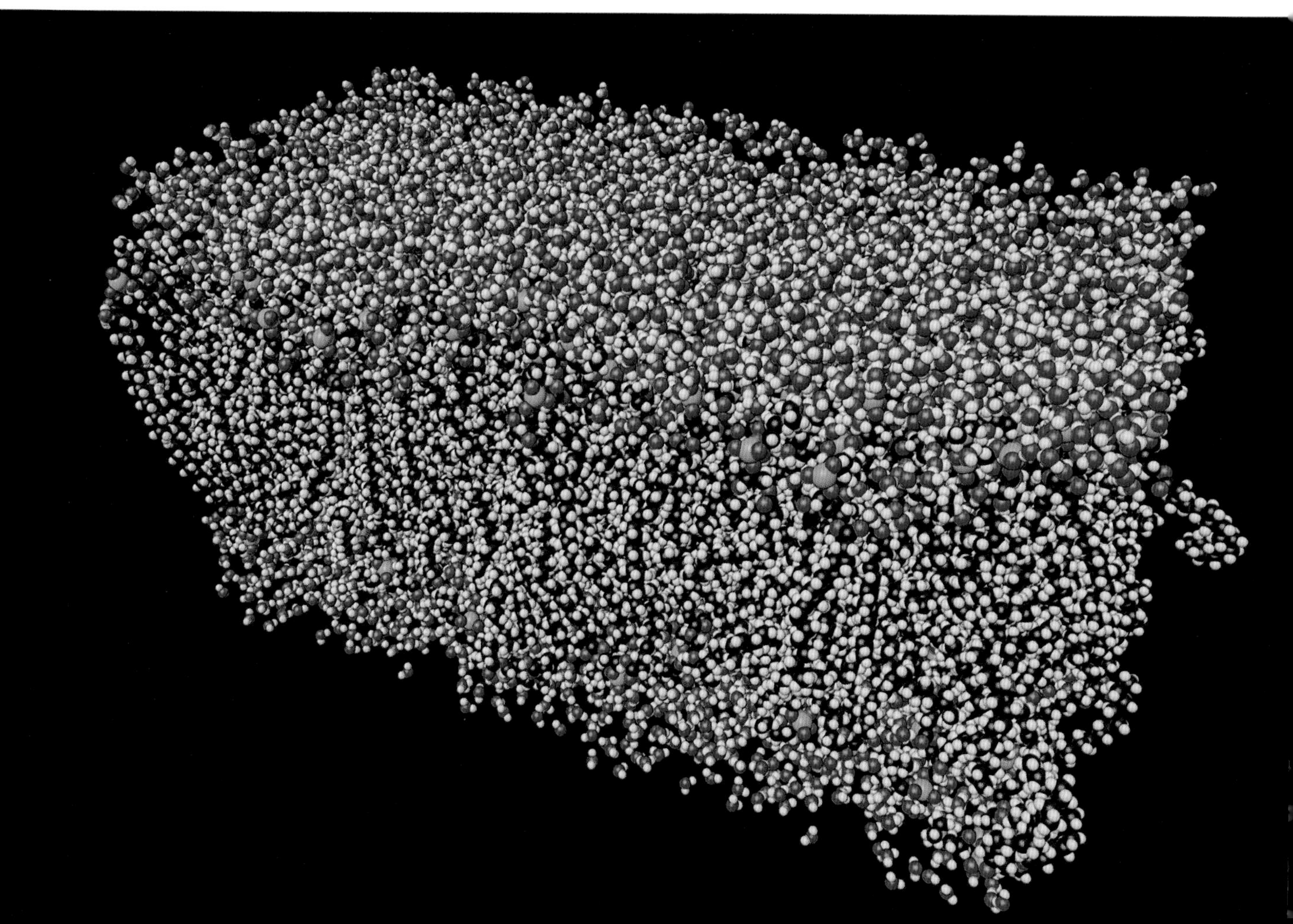

in all cells, the energy stored in carbohydrates is released, to run the cellular processes. Water is crucial in all of this activity—it is there not just to hold the compounds involved in these processes in solution or in suspension—and between 50 and 80 percent of the mass of a biological cell is water, depending on the type of cell. (And since the water molecule is among the smallest and lightest of the molecules present, it accounts for a much higher proportion, when measured as a percentage of the cell's molecular population.)

A small section of the lipid bilayer of the cell membrane. A few layers of water molecules can be seen above and below the membrane. *Source:* Atomic-scale positional data from D. Peter Tieleman, Jason Breed, Herman J. C. Berendsen and Mark S. P. Sansom, "Alamethicin Channels in a Membrane: Molecular Dynamics Simulations," *Faraday Discussions* (1999). https://doi.org/10.1039/a806266h.

The liquid contents of the cell—the *intracellular* water plus all the ions and molecules it carries—are called the *cytoplasm*. In addition to the watery cytoplasm inside, each cell is also surrounded by water, or water-based solutions. This is essential, as cells need to be able to take in certain substances and secrete others. The barrier between the cytoplasm and the *extracellular* water (outside the cell) is the cell membrane. It is made of *phospholipids*, compounds whose molecules have a hydrophilic head (a phosphate group) and a hydrophobic tail—or, rather, two hydrophobic tails (lipids). So phospholipids are amphiphilic, like stearate ions (in soap—see chapter 5). Water molecules herd phospholipid molecules: hydrogen bonds pull the water molecules together, minimizing contact with the phospholipids' hydrophobic tails, while attaching to the hydrophilic head. As a result of these forces, the phospholipid molecules self-assemble into a stable double-layer sheet, or *lipid bilayer*, with their hydrophobic tails tucked firmly into the central part, facing away from both the intracellular and extracellular water. If it were not for water, the lipid bilayer—the cell's defining boundary—would not form.

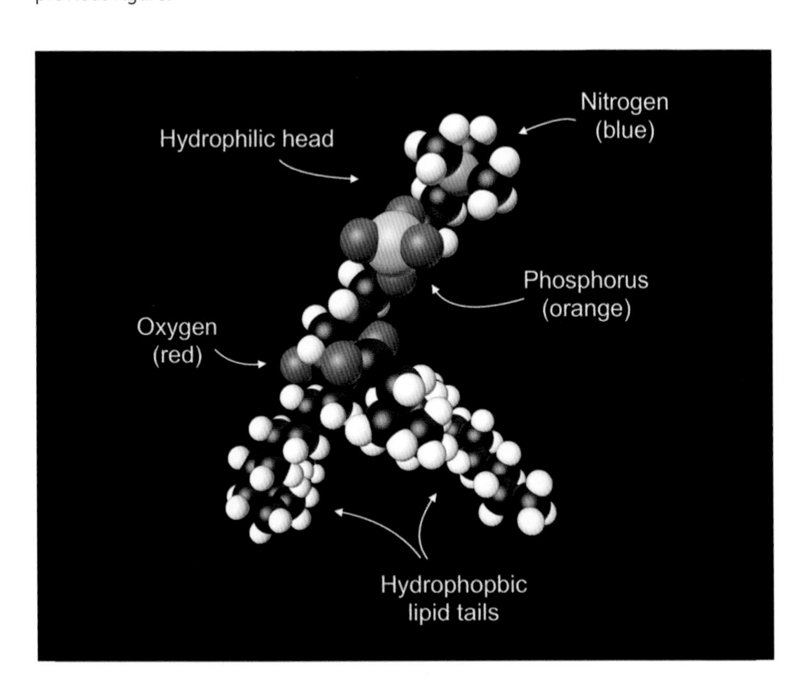

Single molecule of 1-palmitoyl-2-oleoyl-sn-glycero-3-phosphocholine (known as POPC), the phospholipid present in the membrane shown in the previous figure.

The precise nature of intracellular water is a hotly debated topic. There is a huge variety of compounds dissolved in it, and the hydration shells around these solutes affect the network of hydrogen bonds within the water. There are also many interfaces inside the cell, between intracellular water and other substances—including large biomolecules such as proteins and DNA. The crowded nature of the water is likely to have a profound effect on the structure and properties of the intracellular water. In the past, researchers considered the water adjacent to membranes, and the water surrounding large biomolecules, to be a separate phase, distinct from free, bulk water. This water was referred to as *biological water*, and its water molecules were thought to be more or less frozen in place. Research over the past decade or so has ruled out that idea,[1] but the water next to biomolecules is still of vital importance in living processes, and studying it is a major preoccupation of chemists and molecular biologists alike.[2]

Protein molecules in the cytoplasm have both hydrophobic and hydrophilic regions, and the resulting interaction with water plays an important role in their function. Proteins are made of smaller molecules, called amino acids, joined together in long chains, and some amino acids are more hydrophobic than others. Water molecules bind very well to the hydrophilic parts of the molecule, which are electrically charged or polar, and not at all well to those parts that are hydrophobic. The hydrophobic parts end up nudged together, like the lipids in the cell membrane, while the hydrophilic parts remain exposed to the surrounding water. The hydration shells around the hydrophilic parts of protein molecules may attach so firmly that they can be considered part of the molecule's shape. Water molecules switch places much less frequently than normal at hydrophilic regions of proteins, remaining in place for periods thousands of times longer than water molecules normally attach to hydrophilic molecules. Where two or more biomolecules are very close together—as they often are in the crowded cytoplasm—the hydration shells of the molecules may overlap and interact. There is evidence that protons jump from protein to protein through the hydrogen bond network of the intervening water, by the proton-hopping

Grothuss mechanism[3] (see chapter 3), and that this interaction may be crucial to the functioning of the proteins involved. The water around a DNA molecule is also vital to that molecule's function. Water molecules binding to polar sites along the length of a DNA molecule maintain its double helix shape. Genetic information is stored in the sequence of molecules called bases along the length of the molecule. Pairs of bases join together to form the "rungs" in the double-stranded DNA ladder—and the bases are hydrophobic, and therefore held together by the water molecules that surround them.[4] Enzymes that "unzip" the double strand, an essential part of DNA duplication but also the starting point for manufacturing proteins in the cell, have hydrophobic sections in their molecules. As they approach the DNA molecule, they loosen the hydrophobic effect that holds the bases together.

Passing Through

The cell membrane is a very effective barrier between the cell's internal environment and the extracellular space. But a cell cannot survive without exchanging ions, proteins, fats, carbohydrates and other compounds between the intracellular and extracellular water. Some small, nonpolar molecules, such as oxygen and carbon dioxide, can pass fairly freely through the membrane. All solutes *diffuse* (spread out randomly) in a solution—this is why, if you wait long enough, a water-soluble dye will eventually spread out evenly in a bucket of water. As a result of diffusion, solute particles have an overall movement from higher to lower concentration. Since oxygen and carbon dioxide can pass through the membrane freely, they do so. Inside cells in your body right now, oxygen dissolved in (watery) blood plasma is diffusing through the cell membrane, to the (watery) cytoplasm of cells—and carbon dioxide, produced inside the cells, is diffusing in the other direction.

Unlike these nonpolar solutes, polar solutes and ions in the intracellular and extracellular water cannot diffuse through the lipid bilayer, however great the concentration gradient across the membrane. Large polar or partly polar molecules, such as proteins, are instead transported across the membrane, with a small amount of attendant water, inside capsules called *vesicles*. These capsules are bounded by a lipid bilayer, just like the membrane itself. The fact that vesicles are enclosed by a lipid bilayer means they can merge with the cell membrane and open out into the extracellular space, releasing the molecule to dissolve into the extracellular water. The membrane of the used vesicle becomes incorporated into the cell membrane. Neurotransmitters—compounds that pass from one neuron to another to modulate nerve signals—are a good example of molecules that are trafficked via vesicles. They are released into the watery gap between two neurons, called a synapse. A neurotransmitter bursts out of a vesicle at the membrane of the transmitting neuron, passes in solution across to an adjacent neuron, which absorbs the neurotransmitter through its membrane, by the formation of a vesicle that is formed from a piece of lipid bilayer pinched off from the receiving cell's membrane.

Some proteins, and other dissolved substances, can pass into and out of the cell via channels threaded through protein molecules that straddle the membrane. Each channel allows specific

A protein molecule cannot pass through the cell membrane directly. Instead, it is carried through stealthily, inside a membrane-bound sphere called a vesicle.

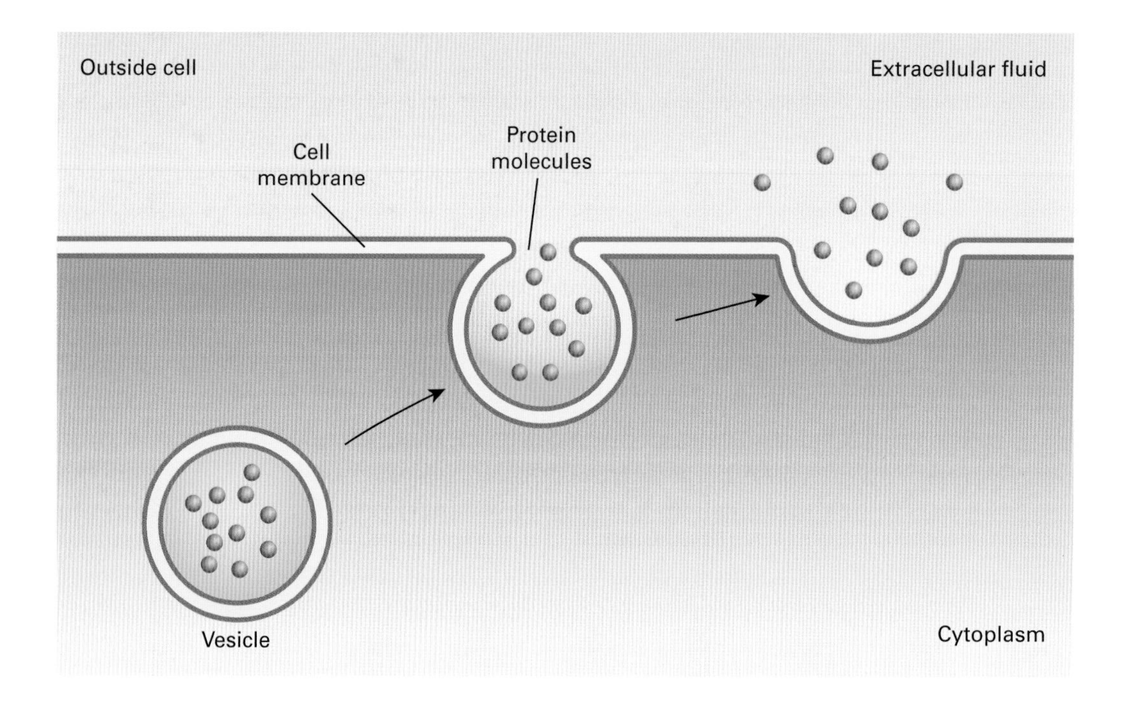

ions or compounds to pass. One very important membrane channel is the potassium ion channel, which is found in many types of cell, but is particularly important in neurons, where the flow of potassium in and out, along the length of the cell, constitutes a nerve signal. At the point where potassium ions enter, the channel opening perfectly fits the first hydration shell of the potassium ion, which contains six water molecules. Potassium ions dissolve extremely well in water, so the channel has its work cut out to bring the ions across the membrane without water molecules attached. It is also vital that no other ions pass through—in particular, sodium ions, which are of a very similar size (and identical electric charge). The shape of the channel, along with the distribution of electric charges within it, mimic potassium's hydration shell, making it favorable for the ions to bind, and then pass through.[5] Several million potassium ions per second can pass through the membrane in this way, first being dehydrated, then rehydrated at the other side. There are several kinds of potassium channels. Most are passive: they simply enable diffusion to occur across the membrane. Most other channels, for other ions, are also passive—but some are active, which means they can move solutes from low to high concentration (this requires an input of energy).

Of course, water itself has to pass in and out of the cell, too. It does so via *osmosis*—a process in which a solvent (in this case, water) can cross a semi-permeable barrier, but the solute particles (ions or molecules) cannot. When there is a concentration gradient across the barrier, the solvent moves from the lower to the higher concentration of solute particles. The cell membrane is a semi-permeable barrier, but it is not ideal, since it does not allow water to pass freely: water is made of polar molecules, so it is more or less immiscible with the lipids in the central, hydrophobic region of the membrane. Only about one in a million water molecules can dissolve there—and since there are so many water molecules present on both sides of the membrane, some do pass through. But osmosis across the membrane is slow and inefficient. Fortunately, the cell membrane of most cells is punctuated by channels that allow water to pass freely across it, allowing osmosis to occur quickly and efficiently.

Leaky Cells

The channels that sit in the membrane and allow water to leak into and out of a cell lie at the center of large protein molecules called *aquaporins*, which sit in the membrane. The channel forms a conduit between intra- and extracellular water. Each aquaporin channel can permit around a billion water molecules to pass through every second—in single file. Aquaporin channels are selective: they allow only water molecules through: ions and other solutes cannot pass. (There is a class of aquaporins, known as *aquaglyceroporins*, however, with larger channels that allow certain solutes through, as well as water.) The aquaporin molecule has conical ends, which funnel water into its constricted middle section. The very narrowest part is 2.8 Å in diameter—equal to the van der Waals radius of the water molecule. The constricted section has electric charges positioned in just the right places to cause each water molecule to tumble through, docking repeatedly to a negative, then a positive, then a negative region. Aquaporins occur in clumps of four identical molecules, each providing a channel through which water can flow.

All the known mammal aquaporins are found in the eye,[6] and over half are also found in the membranes of cells in the kidneys, where they are responsible for reabsorbing water from urine. Aquaporins typically remain open at all times, simply allowing water in or out of the cell as osmosis dictates. That is not to say that the passage of water across the membrane cannot be regulated: it can be, by the variable expression of aquaporins (how many of them are produced inside the cell) and their recruitment (sometimes they already exist inside the cytoplasm, ready to be lifted, when needed, into the membrane). The expression and recruitment of aquaporins can be affected by hormones. A good example of this is vasopressin, which is released from cells in the pituitary gland in response to high salt content in the blood (which the brain registers as thirst). When vasopressin reaches cells in the kidneys, it increases their expression of aquaporin-2 (AQ2), and so elevates the rate at which water can be reabsorbed.[7] Several aquaporins are up-regulated (their expression increased)

Three different views of an aquaporin protein channel. Above, two models of the aquaporin are shown embedded in the cell membrane; the right-hand view shows the protein's secondary structure (helices) and tertiary structure (overall shape). Below is a top view showing how the individual aquaporin proteins are arranged in a "tetramer"—four identical proteins attached together.

in various ways during pregnancy, largely to build up amniotic fluid in the uterus.[8] Aquaporins in the cells in the epithelium (lining) of the intestines absorb large amounts of water, from partially digested food. Certain pathogens, or disease-causing agents, including some strains of *Escherichia coli*, release toxins that adversely affect the expression of these aquaporins. With the population of aquaporins dramatically reduced, much of the water remains inside the intestines.[9] The diarrhea that ensues is advantageous to the bacteria, as it allows them to be expelled in water, so they may pass to another host.

Aquaporins are also common in fungi and in plants.[10] Here, they are more likely to allow certain solutes through, as well as water, as plants and fungi need nutrients in the water they absorb. Unlike the aquaporins in animal cells, some aquaporins in plants have mechanisms by which they can close—in response to drought, for example.[11] Aquaporins also have a role to play in some of the more

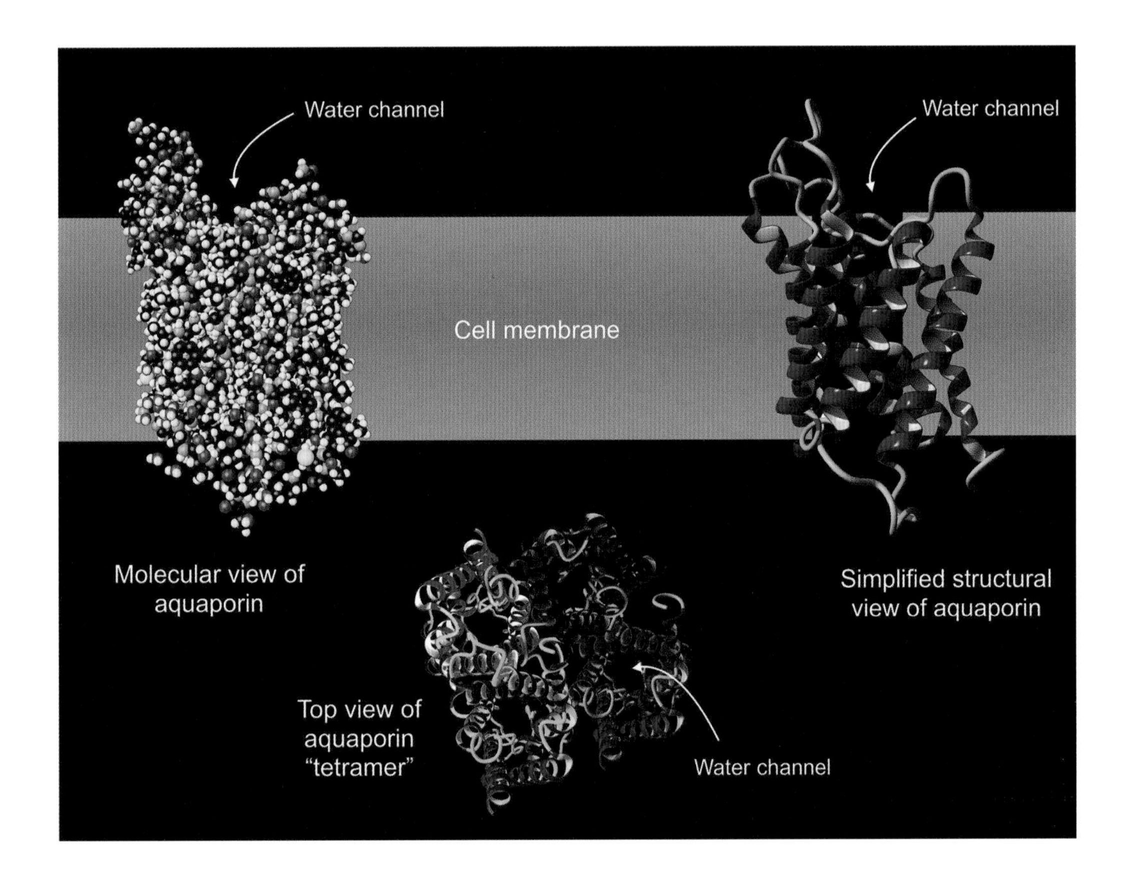

exciting plant-based phenomena. For example, rapidly opening and closing aquaporins are present in cells of the Venus fly trap plant, where they enable the buildup of water pressure—which causes the sudden closure of the plant's deadly lobes when an insect lands.[12] Not all aquaporins sit in the cell's membrane; some are found in the membranes of organelles within the cell. One organelle that has its own membrane, and its own aquaporins, is the water-filled vacuole, which is important in maintaining *turgor* (rigidity) in plant cells. Aquaporin-mediated turgor in plants can generate pressures tens or hundreds of times greater than atmospheric pressure, which can be needed, for example, when a germinating seed needs to break through its coat.[13]

From Roots to Leaves

Aquaporins are present on the membrane of root hair cells, whose job is to absorb water (and nutrients) from the soil. The membrane of a root cell has a large surface area, with a large number of aquaporins—so these cells are well adapted to taking in water. A high concentration of dissolved substances in the water inside the roots means that water from outside will flow across the membrane, by osmosis. This creates a pressure, which helps push water up toward the leaves. However, aquaporins are not otherwise involved in transporting water from the roots up to the leaves. This takes place in specialized tissue called *xylem*, in a plant's stem or trunk. Xylem cells are empty dead cells with rigid cell walls. There is no cell wall, however, at the top or the bottom of each cell, so the dead cells form a hollow continuous pipe. In addition to the push created by osmosis at the roots, water is also pulled from above. The evaporation of water through pores called *stomata* in the leaves creates a negative pressure, pulling more water up through the xylem—this is called *transpirational pull*. Water's strong cohesion prevents the column of water from breaking under this negative pressure. Occasionally, the column does break, forming a vapor inside the xylem tube—a process called *cavitation*, which can be catastrophic for a plant.[14]

Cavitation aside, the combined effects of osmotic pressure, transpirational pull, and cohesion—as well as capillary action—can lift water hundreds of feet in a tall tree. Giant redwoods, for example, can grow to more than 300 feet, and their xylem is always full of water. The amounts of water a tree can suck out of the soil is impressive: a large, mature oak tree can drain the soil of 40,000 gallons of water each year.

Of course, a plant does not drag all that water up from the ground just to have it pass out through the leaves' stomata. There is another reason why plants suck up water. Some of it is destined to be processed to make fuel that powers the set of chemical reactions that keep the plant alive: photosynthesis. In plants, photosynthesis takes place in membrane-bound organelles called *chloroplasts.* (Plants are not the only organisms in which photosynthesis takes place: algae and certain single-celled bacteria, called cyanobacteria, also photosynthesize.)

At the heart of photosynthesis is a process in which the energy of sunlight splits water molecules. Water is a great choice as a raw material for photosynthesis—and not only because it is plentiful. It takes a great deal of energy to break apart a water molecule, so splitting water enables plants to store large amounts of energy. In the first stage of photosynthesis, molecules of the green pigment chlorophyll in a large complex of molecules called *photosystem II* capture the light energy and use it to boost the energy of its electrons. Photosystem II sits in the membrane of a compartment inside the chloroplast, called the *thylakoid.* Some of the energy of the electrons is used to pump hydrogen ions (H^+) from water on one side of the thylakoid membrane to the other. The resulting increase in hydrogen ion concentration on one side of the membrane is crucial, as we shall see. The electrons lost from photosystem II have to be replaced—and they come from water molecules. One part of photosystem II, called the *oxygen-evolving center* (or the *water-splitting center*), steals electrons from water molecules—and since electrons are what bind hydrogen and oxygen together, the molecules fall apart. The oxygen-evolving complex takes two electrons per water molecule. For every two molecules ($2H_2O$), the plant yields four electrons ($4e^-$),

four hydrogen ions ($4H^+$) and two oxygen atoms (2O). The oxygen atoms immediately combine to form an oxygen molecule (O_2), the waste product of the process. The hydrogen ions released in the destruction of the water molecules remain in solution, on the same side of the membrane as the hydrogen ions produced earlier. The high concentration of hydrogen ions on that side of the thylakoid membrane drives the next step. Hydrogen ions pass through a channel in a molecule called *ATP synthase*, which sits on the thylakoid membrane. The energy of the hydrogen ions

▲ **Photosystem II**—the complex structure that is responsible for capturing the energy of sunlight and breaking apart water molecules. The location of the oxygen-evolving complex, or water-splitting complex, is highlighted.

passing through ATP synthase is used to build molecules of the high-energy compound ATP (adenosine triphosphate), which is the universal energy currency within all cells, and therefore for all life.

Building up and Breaking Down

Still inside the chloroplast, the energy of ATP is used to manufacture high-energy carbohydrate compounds containing three carbon atoms each, from carbon dioxide molecules. These three-carbon molecules form the basis of simple sugar molecules (monosaccharides), as well as amino acids (the building blocks of proteins) and fats. One of the most important monosaccharides, glucose, acts as a convenient energy storage molecule in its own right: in *respiration*—the reverse of photosynthesis—glucose is broken down, releasing energy that builds ATP, and that ATP is used to drive a host of processes inside the cell. Another product of respiration is water—the metabolic water mentioned in chapter 2. For each molecule of glucose that combines with oxygen, six molecules of water are produced. Glucose and other monosaccharides double up to make disaccharides, such as sucrose—and join in larger numbers to make polysaccharides, such as starch (a longer-term form of energy storage) and cellulose (which plants use to build their cell walls). Glucose, sucrose and the other products of photosynthesis pass out of the chloroplast and into a separate system of water-filled tubes called the *phloem*, which distributes these precious resources around the plant.

The reaction that builds larger carbohydrates from monosaccharide building blocks is called a *condensation reaction*, because it produces a molecule of water. There are many other condensation reactions at work inside cells, including the formation of proteins from their amino acid building blocks. The reverse reaction is called *hydrolysis* (literally, "splitting with water"). Molecules of ATP break down by hydrolysis, for example: the lone pair of electrons on the oxygen atom of a nearby water molecule affects the electrons in a group of atoms—a *phosphate* group—at one end of the ATP molecule. This causes the bond between that phosphate group and the rest of the molecule to break, and it is this loss of the phosphate group that turns ATP into ADP (adenosine diphosphate), releasing energy. The water molecule is broken apart as a result, releasing a hydrogen ion into the solution and attaching a hydroxide group to the phosphate group. Each day, the reactions inside the cells of your body form and break down more than your body weight of water molecules.

Micrograph of a cross-section through the stem of a maize plant (genus *Zea*), showing "vascular bundles" made up of (larger) xylem tubes and phloem tubes, surrounded by a ring of (darker stained) supportive cells.

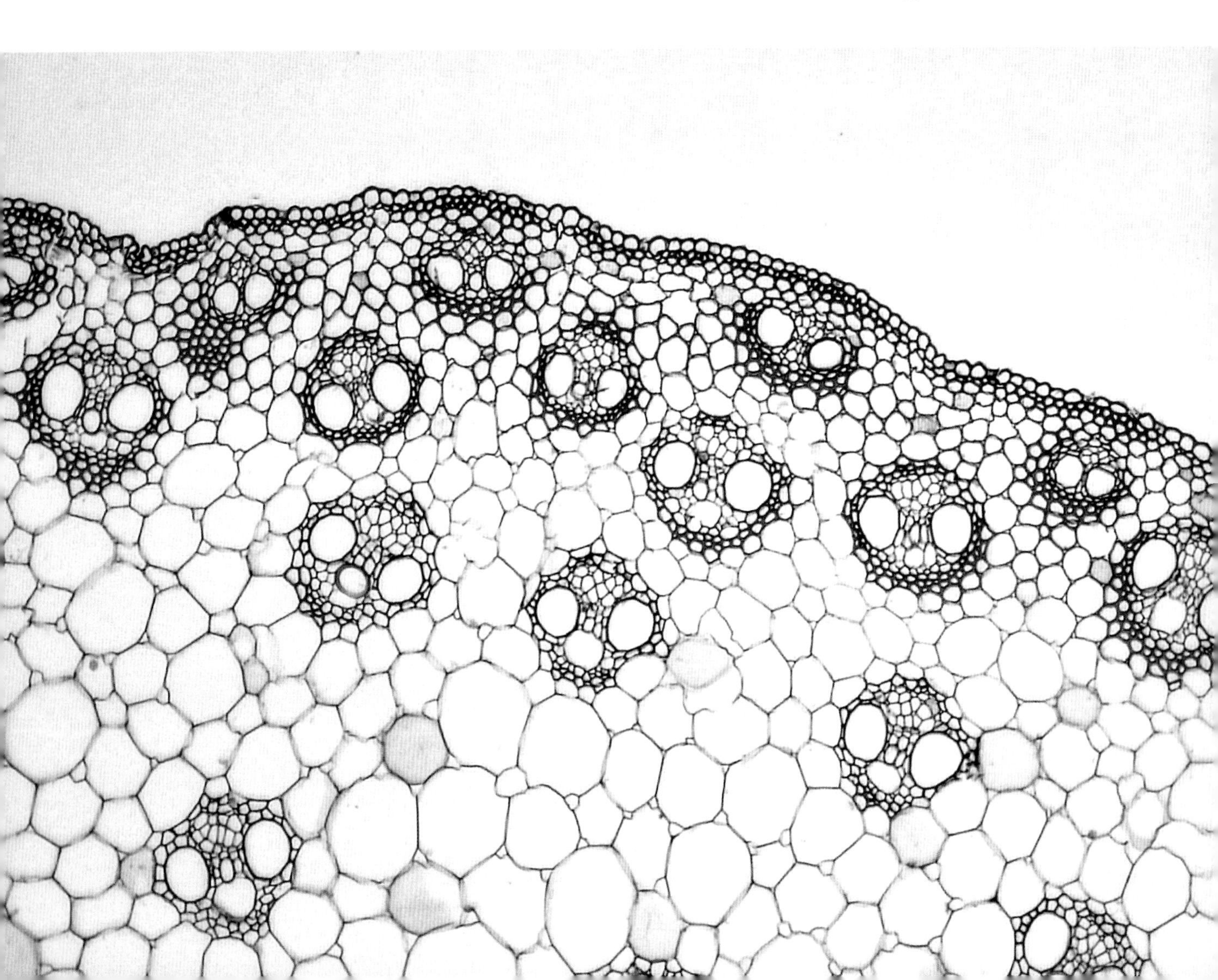

Example of a condensation reaction. Glucose and fructose, two monosaccharides, join to form sucrose, a disaccharide. The other product of the reaction is one molecule of water.

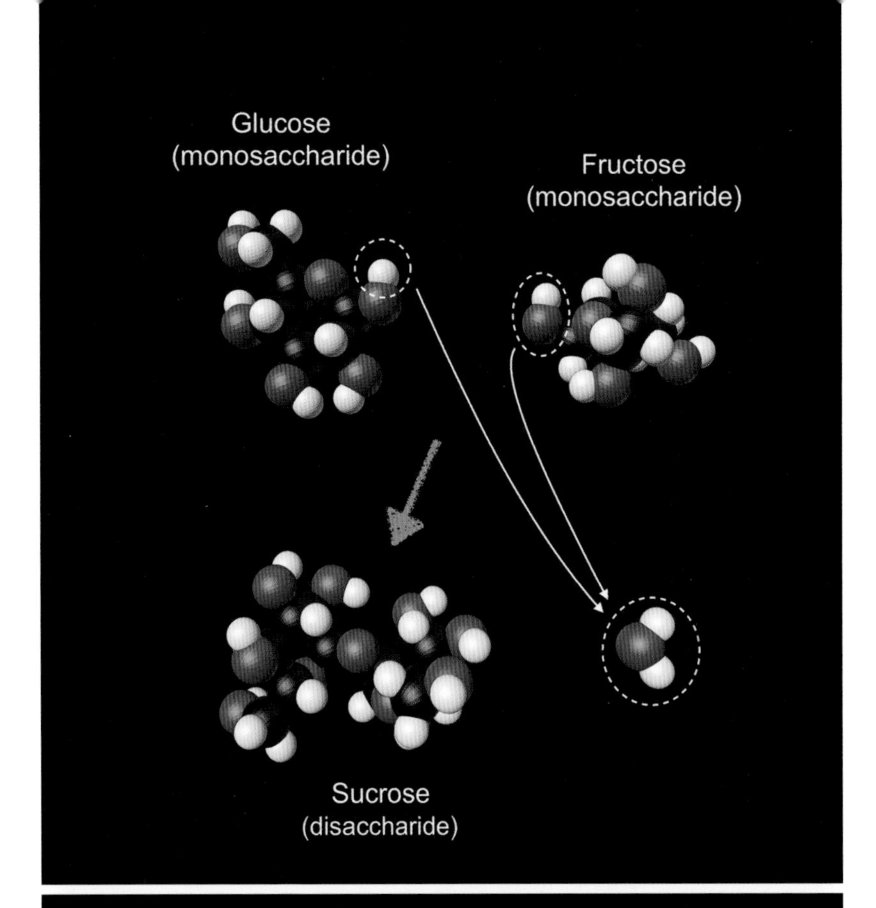

Example of a hydrolysis reaction: ATP (adenosine triphosphate) to ADP (adenosine diphosphate). The partial negative charge of the lone pair (yellow) on an oxygen atom in a water molecule creates instability in the bond between the last phosphate group and the rest of the ATP molecule. As a result, the phosphate group breaks away, leaving ADP. The water molecule splits into H^+ and OH^-. The OH^- becomes part of the freed phosphate group, while the H^+ joins another water molecule, to form a hydronium ion.

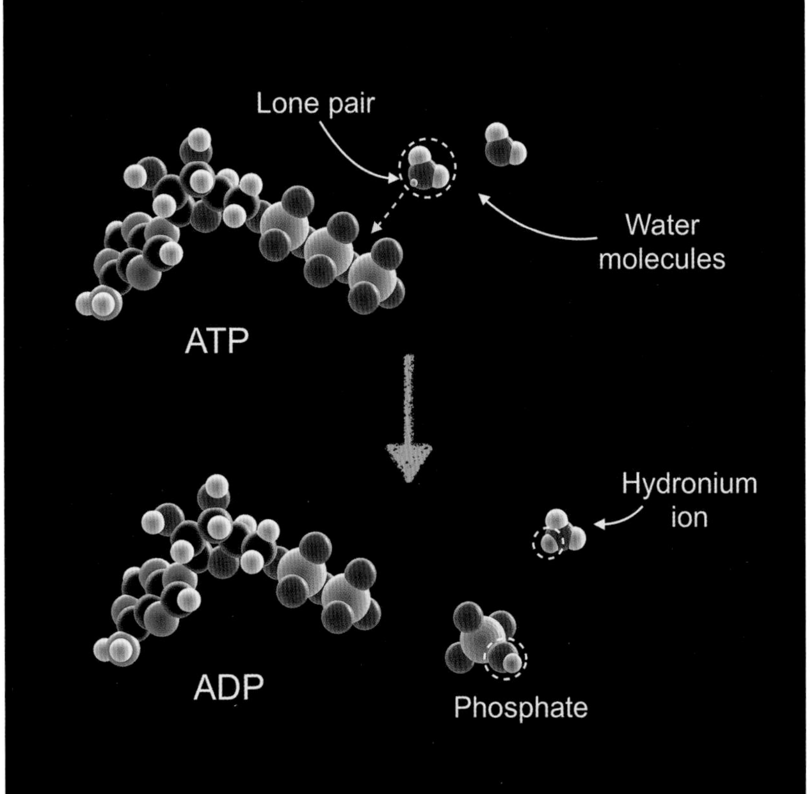

A "Warm Little Pond"

Water's ability to hold so many substances in solution or suspension—so that they can react together—is vital in the reactions that make life possible. And water is intimately involved in all those reactions, in which biomolecules allow cells to survive and thrive. It is no surprise, then, that evolutionary biologists suggest—almost assume—that life began in water. This idea dates back at least as far as 1871, when English naturalist Charles Darwin conceived in a letter to his friend Joseph Hooker a "warm little pond" containing "ammonia and phosphoric salts," in which a protein could form that would then "undergo still more complex changes."

The first living things on our planet appeared at least 3.5 billion years ago. Convention has it that life originated in the oceans—most likely in the heat and chemical chaos of hydrothermal vents at the seafloor. But the ratio of sodium to potassium ions in seawater does not match that same ratio in living cells. This leads some to suggest that it was in shallow freshwater lakes or ponds on land that life first appeared. Another reason for suggesting that shallow water on land would have been the right environment to kickstart life is that nitrogen, a crucial element in living systems, would have been more available there. In 2019, researchers at MIT published research that backs up this idea.[15] The researchers carried out an analysis of nitrogen sources and sinks in water on the early Earth. Nitrogen (N_2) is plentiful in the atmosphere, but it is a very stable molecule: it does not dissolve, and is not easily available to take part in chemical reactions. Nitrogen oxides do dissolve in water, and provide a way in which nitrogen atoms become available. Lightning creates nitrogen oxides in the air, which dissolve in water. But in the sea, they may have become too dilute to have any chance of forming those first biomolecules—and reactions with dissolved iron would have sent the nitrogen back into the air. Another problem with life originating in the oceans concerns phosphorylation—the reaction in which a phosphate group is added to another molecule, as when ADP becomes ATP. Today, phosphorylation is aided by enzymes (proteins that act as catalysts). When life first evolved, it would have to have happened spontaneously—but that happens more favorably at surfaces than in bulk liquid. In a 2017 study, researchers observed plentiful phosphorylation in tiny droplets of water, leading them to speculate that life may have begun not in the oceans or ponds, but rather in tiny micro-droplets of water.[16]

Wherever life began, it must have relied on water just as much as living things do today. As far as we know, life existed solely in water for at least its first 2.5 billion years. Lichen—a symbiotic marriage of fungi and algae—began to colonize the land around 700 million years ago. The fungi were good at finding water, while the algae provided energy, by photosynthesis. Lichens had a profound effect on the land: at first, the landscape was hard and rocky, but lichens broke down the rocks, making soil that would hold water—a necessary prerequisite for the next colonizers of the land. Around 430 million years ago the first plants began thriving on land. In order to establish themselves away from a constant supply of water, early plants had to develop spores (the equivalent of seeds) that could survive out of water. As plants began to thrive on land, competition for light meant they evolved to be taller. That nudged evolution into producing the first *vascular* ("vesseled") plants, which could carry water from the soil up to the leaves. By 300 million years ago, during the Carboniferous period, enormous trees up to 100 feet tall dominated landscapes across the world.

Animals could colonize the land only if they could find a way to keep the insides of their bodies wet at all times; given the importance of water, desiccation is not an option. So they evolved waterproof skins or exoskeletons to keep the water from escaping. They also had to have ways to reproduce on land without water. Amphibians simply returned to the water, but reptiles evolved eggs with rigid shells, which came with their own water supply. In mammals, of course, that amniotic fluid bathes the developing fetus. Animals also had to be able to take in oxygen and excrete carbon dioxide, which for an animal underwater would diffuse directly to and from the water. Early amphibians evolved moist membranes that were exposed to the air, so that gas exchange could take place—we call them lungs. All of these adaptations are still in play today, since the fundamental need for water is still as important as ever.

In 1971, Albert Szent-Györgyi, quoted at the beginning of this chapter, wrote "Biology has forgotten water, or never discovered it."[17] That is certainly not true today: researchers across many disciplines of science recognize the central role water plays in life, and have made serious inroads to understand its important, complex subtleties. Water is a simple molecule, and one of the smallest. But its unique set of properties make it ideal for enabling and sustaining the living world. We owe our life to it.

Notes

Chapter 1

1. N. Laporte, R. S. Ellis, F. Boone, F. E. Bauer, D. Quénard, G. W. Roberts-Borsani et al., "Dust in the Reionization Era: ALMA Observations of a $z = 8.38$ Gravitationally Lensed Galaxy," *The Astrophysical Journal* (2017). https://doi.org/10.3847/2041-8213/aa62aa.

2. F. Dulieu, L. Amiaud, E. Congiu, J. H. Fillion, E. Matar, A. Momeni et al., "Experimental Evidence for Water Formation on Interstellar Dust Grains by Hydrogen and Oxygen Atoms," *Astronomy and Astrophysics* (2010). https://doi.org/10.1051/0004-6361/200912079.

3. "Revealing the Molecular Universe: One Antenna Is Never Enough," in ASP Conference Series, vol. 356, *Proceedings of the Conference Held 9–10 September, 2005, at University of California, Berkeley, California, USA*, ed. D. C. Backer, J. W. Moran and J. L. Turner (San Francisco: Astronomical Society of the Pacific, 2006), 81.

4. C. M. Bradford, A. D. Bolatto, P. R. Maloney, J. E. Aguirre, J. J. Bock, J. Glenn et al., "The Water Vapor Spectrum of APM 08279+5255," *ArXiv E-Prints* (2011).

5. P. Sonnentrucker, D. A. Neufeld, T. G. Phillips, M. Gerin, D. C. Lis, M. De Luca et al., "Detection of Hydrogen Fluoride Absorption in Diffuse Molecular Clouds with *Herschel*/HIFI: An Ubiquitous Tracer of Molecular Gas," *Astronomy and Astrophysics* (2010). https://doi.org/10.1051/0004-6361/201015082.

6. Michael J. Kaufman and David A. Neufeld, "Water Maser Emission from Magnetohydrodynamic Shock Waves," *The Astrophysical Journal* (1996). https://doi.org/10.1086/176645.

7. Jonathan Tennyson and Oleg L. Polyansky, "Water on the Sun: The Sun Yields More Secrets to Spectroscopy," *Contemporary Physics* 39:4 (1998): 283–294.

8. Alexandra Witze, "Second-Ever Interstellar Comet Contains Alien Water," *Nature* (2019). https://doi.org/10.1038/d41586-019-03334-5.

9. M. Ali-Dib, O. Mousis, J.-M. Petit and J. I. Lunne, "Resolving the Inconsistency between the Ice Giants and Cometary D/H Ratios," in *SF2A-2014: Proceedings of the Annual Meeting of the French Society of Astronomy and Astrophysics*, ed. J. Ballet, F. Martins, F. Bournaud, R. Monier and C. Reylé (Paris: French Society of Astronomy & Astrophysics, 2014), 169–172.

10. Kim Reh, Mark Hofstadter, John Elliott and Amy Simon, "On to the Ice Giants; Pre-Decadal Study Summary," European Geophysical Union, 2017.

11. Marius Millot, Federica Coppari, J. Ryan Rygg, Antonio Correa Barrios, Sebastien Hamel, Damian C. Swift et al., "Nanosecond X-Ray Diffraction of Shock-Compressed Superionic Water Ice," *Nature* (2019). https://doi.org/10.1038/s41586-019-1114-6.

12. G. L. Bjoraker, M. H. Wong, I. de Pater, T. Hewagama, M. Ádámkovics and G. S. Orton, "The Gas Composition and Deep Cloud Structure of Jupiter's Great Red Spot," *The Astronomical Journal* (2018). https://doi.org/10.3847/1538-3881/aad186.

13. R. G. Martin and M. Livio, "On the Evolution of the Snow Line in Protoplanetary Discs," in *Proceedings of the International Astronomical Union* (2013). https://doi.org/10.1017/S1743921313008235.

14. O. Ruesch, T. Platz, P. Schenk, L. A. McFadden, J. C. Castillo-Rogez, L. C. Quick et al., "Cryovolcanism on Ceres," *Science* (2016). https://doi.org/10.1126/science.aaf4286.

15. Z. Jin and M. Bose, " New Clues to Ancient Water on Itokawa," *Science Advances* (2019). https://doi.org/10.1126/sciadv.aav8106.

16. Cheng Zhu, Parker B. Crandall, Jeffrey J. Gillis-Davis, Hope A. Ishii, John P. Bradley, Laura M. Corley et al., "Untangling the Formation and Liberation of Water in the Lunar Regolith," *Proceedings of the National Academy of Sciences of the United States of America* (2019). https://doi.org/10.1073/pnas.1819600116.

17. Shuai Li, Paul G. Lucey, Ralph E. Milliken, Paul O. Hayne, Elizabeth Fisher, Jean Pierre Williams et al., "Direct Evidence of Surface Exposed Water Ice in the Lunar Polar Regions," *Proceedings of the National Academy of Sciences of the United States of America* (2018). https://doi.org/10.1073/pnas.1802345115.

18. S. Nerozzi and J. W. Holt, "Buried Ice and Sand Caps at the North Pole of Mars: Revealing a Record of Climate Change in the Cavi Unit with SHARAD," *Geophysical Research Letters* (2019). https://doi.org/10.1029/2019GL082114.

19. Sylvain Piqueux, Jennifer Buz, Christopher S. Edwards, Joshua L. Bandfield, Armin Kleinböhl, David M. Kass et al., "Widespread Shallow Water Ice on Mars at High Latitudes and Midlatitudes," *Geophysical Research Letters* (2019). https://doi.org/10.1029/2019GL083947.

20. R. Orosei, S. E. Lauro, E. Pettinelli, A. Cicchetti, M. Coradini, B. Cosciotti et al., "Radar Evidence of Subglacial Liquid Water on Mars," *Science* (2018). https://doi.org/10.1126/science.aar7268.

21. W. Rapin, B. L. Ehlmann, G. Dromart, J. Schieber, N. H. Thomas, W. W. Fischer et al., "An Interval of High Salinity in Ancient Gale Crater Lake on Mars," *Nature Geoscience* (2019). https://doi.org/10.1038/s41561-019-0458-8.

22. Francesco Salese, Monica Pondrelli, Alicia Neeseman, Gene Schmidt and Gian Gabriele Ori, "Geological Evidence of Planet-Wide Groundwater System on Mars," *Journal of Geophysical Research: Planets* (2019). https://doi.org/10.1029/2018JE005802.

23. G. L. Villanueva, M. J. Mumma, R. E. Novak, H. U. Käufl, P. Hartogh, T. Encrenaz et al., "Strong Water Isotopic Anomalies in the Martian Atmosphere: Probing Current and Ancient Reservoirs," *Science* (2015). https://doi.org/10.1126/science.aaa3630.

24. A. T. Basilevsky, E. V. Shalygin, D. V. Titov, W. J. Markiewicz, F. Scholten, T. Roatsch et al., "Geologic Interpretation of the Near-Infrared Images of the Surface Taken by the Venus Monitoring Camera, Venus Express," *Icarus* (2012). https://doi.org/10.1016/j.icarus.2011.11.003.

25. M. J. Way, Anthony D. Del Genio, Nancy Y. Kiang, Linda E. Sohl, David H. Grinspoon, Igor Aleinov et al., "Was Venus the First Habitable World of Our Solar System?," *Geophysical Research Letters* (2016). https://doi.org/10.1002/2016GL069790.

Chapter 2

1. J. Leconte, F. Forget, B. Charnay, R. Wordsworth and A. Pottier, "Increased Insolation Threshold for Runaway Greenhouse Processes on Earth-Like Planets," *Nature* (2013). https://doi.org/10.1038/nature12827.

2. Hideyuki Nakano, Naoki Hirakawa, Yasuhiro Matsubara, Shigeru Yamashita, Takuo Okuchi, Kenta Asahina et al., "Precometary Organic Matter: A Hidden Reservoir of Water inside the Snow Line," *Scientific Reports* (2020). https://doi.org/10.1038/s41598-020-64815-6.

3. J. Wu, S. J. Desch, L. Schaefer, L. T. Elkins-Tanton, K. Pahlevan and P. R. Buseck, "Origin of Earth's Water: Chondritic Inheritance plus Nebular Ingassing and Storage of Hydrogen in the Core," *Journal of Geophysical Research: Planets* (2018). https://doi.org/10.1029/2018JE005698.

4. D. Deming, "Born to Trouble: Bernard Palissy and the Hydrologic Cycle," *Ground Water* (2005). https://doi.org/10.1111/j.1745-6584.2005.00119.x.

5. Jianping Huang, Jiping Huang, Xiaoyue Liu, Changyu Li, Lei Ding and Haipeng Yu, "The Global Oxygen Budget and Its Future Projection," *Science Bulletin* (2018). https://doi.org/10.1016/j.scib.2018.07.023.

6. David C. Catling and Kevin J. Zahnle, "The Planetary Air Leak: As Earth's Atmosphere Slowly Trickles away into Space, Will Our Planet Come to Look like Venus?," *Scientific American* (May 2009).

7. C. Cai, D. A. Wiens, W. Shen and M. Eimer, "Water Input into the Mariana Subduction Zone Estimated from Ocean-Bottom Seismic Data," *Nature* (2018). https://doi.org/10.1038/s41586-018-0655-4.

8. M. J. Walter, A. R. Thomson, W. Wang, O. T. Lord, J. Ross, S. C. McMahon et al., "The Stability of Hydrous Silicates in Earth's Lower Mantle: Experimental Constraints from the Systems $MgO-SiO_2-H_2O$ and $MgO-Al_2O_3-SiO_2-H_2O$," *Chemical Geology* (2015). https://doi.org/10.1016/j.chemgeo.2015.05.001.

9. Narangoo Purevjav, Takuo Okuchi, Naotaka Tomioka, Jun Abe and Stefanus Harjo, "Hydrogen Site Analysis of Hydrous Ringwoodite in Mantle Transition Zone by Pulsed Neutron Diffraction," *Geophysical Research Letters* (2014). https://doi.org/10.1002/2014GL061448.

10. Kenneth G. Libbrecht, *Snow Crystals* (self-published, 2019). See snowcrystals.com.

11. Peter V. Hobbs and Arthur L. Rangno, "Super-Large Raindrops," *Geophysical Research Letters* (2004). https://doi.org/10.1029/2004GL020167.

12. Massimo Bernardi, Piero Gianolla, Fabio Massimo Petti, Paolo Mietto and Michael J. Benton, "Dinosaur Diversification Linked with the Carnian Pluvial Episode," *Nature Communications* (2018). https://doi.org/10.1038/s41467-018-03996-1.

13. K. Voudouris, M. Valipour, A. Kaiafa, X. Y. Zheng, R. Kumar, K. Zanier et al., "Evolution of Water Wells Focusing on Balkan and Asian Civilizations," *Water Science and Technology: Water Supply* (2019). https://doi.org/10.2166/ws.2018.114.

14. M. Altaweel, "Southern Mesopotamia: Water and the Rise of Urbanism," *Wiley Interdisciplinary Reviews: Water* (2019). https://doi.org/10.1002/wat2.1362.

15. Margaret R. Bunsen, *Encyclopedia of Ancient Egypt* (Gramercy Books, 1999), 277.

16. Saifulla Khan, "Sanitation and Wastewater Technologies in Harappa/Indus Valley Civilization (ca. 2600–1900 BC)," in *Evolution of Sanitation and Wastewater Technologies through the Centuries* (International Water Association, 2014).

17. L. Giosan, W. D. Orsi, M. Coolen, C. Wuchter, A. G. Dunlea, K. Thirumalai et al., "Neoglacial Climate Anomalies and the Harappan Metamorphosis," *Climate of the Past* (2018). https://doi.org/10.5194/cp-14-1669-2018.

18. S. Macrae and G. Iannone, "Understanding Ancient Maya Agricultural Terrace Systems through Lidar and Hydrological Mapping," *Advances in Archaeological Practice* (2016). https://doi.org/10.7183/2326-3768.4.3.371.

19. "Water Conflict," http://www.worldwater.org/water-conflict.

20. Peter Gleick, *The World's Water*, vol. 9 (Pacific Institute, 2018), 141.

21. S. Jasechko and D. Perrone, "Hydraulic Fracturing near Domestic Groundwater Wells," *Proceedings of the National Academy of Sciences of the United States of America* (2017). https://doi.org/10.1073/pnas.1701682114.

22. F. Farinosi, C. Giupponi, A. Reynaud, G. Ceccherini, C. Carmona-Moreno, A. De Roo et al., "An Innovative Approach to the Assessment of Hydro-Political Risk: A Spatially Explicit, Data Driven Indicator of Hydro-Political Issues," *Global Environmental Change* (2018). https://doi.org/10.1016/j.gloenvcha.2018.07.001.

23. washdata.org.

Chapter 3

1. Henry Cavendish, "Three Papers, Containing Experiments on Factitious Air, by the Hon. Henry Cavendish, F. R. S," *Philosophical Transactions (1683–1775)* 56 (1766): 141–184.

2. Sidney M. Edelstein, "Priestley Settles the Water Controversy," *Chymia* (1948). https://doi.org/10.2307/27757119.

3. David Philip Miller, *Discovering Water: James Watt, Henry Cavendish and the Nineteenth-Century "Water Controversy"* (Taylor and Francis, 2004), 28.

4. Henry Guerlac, *Lavoisier—The Crucial Year: The Background and Origin of His First Experiments on Combustion in 1772* (Cornell University Press, 1961).

5. Henry Cavendish, "Experiments on Air. By Henry Cavendish, Esq. F.R.S. and A.S.," *Philosophical Transactions of the Royal Society of London* 75 (1785): 372–384.

6. Antoine Laurent Lavoisier, *Traité élémentaire de chimie t.* 1 (1789) (Chez Cuchet, 1789).

7. R. De Levie, "The Electrolysis of Water," *Journal of Electroanalytical Chemistry* (1999). https://doi.org/10.1016/S0022-0728(99)00365-4.

8. William Nicholson, "Account of the New Electrical or Galvanic Apparatus of Sig. Alex. Volta, and Experiments Performed with the Same," *Journal of Chemistry, Natural Philosophy and the Arts* 4 (1800): 179.

9. John Dalton, *New System of Chemical Philosophy* (1808), 171.

10. Xianyong Wu, Jessica J. Hong, Woochul Shin, Lu Ma, Tongchao Liu, Xuanxuan Bi et al., "Diffusion-Free Grotthuss Topochemistry for High-Rate and Long-Life Proton Batteries," *Nature Energy* (2019). https://doi.org/10.1038/s41560-018-0309-7.

Chapter 4

1. For an overview of methane hydrates, see C. D. Ruppel, "Methane Hydrates and Contemporary Climate Change," *Nature Education Knowledge* (2011). https://www.nature.com/scitable/knowledge/library/methane-hydrates-and-contemporary-climate-change-24314790/.

2. P. S. R. Prasad and B. Sai Kiran, "Clathrate Hydrates of Greenhouse Gases in the Presence of Natural Amino Acids: Storage, Transportation and Separation Applications," *Scientific Reports* (2018). https://doi.org/10.1038/s41598-018-26916-1.

3. Victor F. Petrenko and Robert W. Whitworth, *Physics of Ice* (Oxford University Press, 1999), 10.

4. H. König, "Eine kubische Eismodifikation," *Zeitschrift Für Kristallographie—Crystalline Materials* 105:1–6 (1943): 279–286. https://doi.org/10.1524/zkri.1943.105.1.279.

5. W. X. Zhang, C. He, J. S. Lian and Q. Jiang, "Selected Crystallization of Water as a Function of Size," *Chemical Physics Letters* (2006). https://doi.org/10.1016/j.cplett.2006.01.085.

6. Benjamin J. Murray, Tamsin L. Malkin and Christoph G. Salzmann, "The Crystal Structure of Ice under Mesospheric Conditions," *Journal of Atmospheric and Solar-Terrestrial Physics* (2015). https://doi.org/10.1016/j.jastp.2014.12.005.

7. Percy Bridgman, "General Survey of Certain Results in the Field of High-Pressure Physics," Nobel Lecture, December 11, 1946.

8. A wealth of information about all the various phases of ice is available at Martin Chaplin's website: http://www1.lsbu.ac.uk/water/ice_phases.html.

9. T. Matsui, M. Hirata, T. Yagasaki, M. Matsumoto and H. Tanaka, "Communication: Hypothetical Ultralow-Density Ice Polymorphs," *Journal of Chemical Physics* (2017). https://doi.org/10.1063/1.4994757.

10. E. F. Burton and W. F. Oliver, "X-Ray Diffraction Patterns of Ice [2]," *Nature* (1935). https://doi.org/10.1038/135505b0.

11. O. Mishima, L. D. Calvert and E. Whalley, "'Melting Ice' I at 77 K and 10 kbar: A New Method of Making Amorphous Solids," *Nature* (1984). https://doi.org/10.1038/310393a0.

12. F. Martelli, S. Torquato, N. Giovambattista and R. Car, "Large-Scale Structure and Hyperuniformity of Amorphous Ices," *Physical Review Letters* (2017). https://doi.org/10.1103/PhysRevLett.119.136002.

13. "So Much More to Know . . . ," *Science* 309:5731 (July 1, 2005): 78–102. DOI: 10.1126/science.309.5731.78b.

14. For a thorough review of the history and importance of the RDF for liquid water, see A. K. Soper, "The Radial Distribution Functions of Water as Derived from Radiation Total Scattering Experiments: Is There Anything We Can Say for Sure?," *ISRN Physical Chemistry* (2013). https://doi.org/10.1155/2013/279463.

15. Ivan Brovchenko and Alla Oleinikova, "Multiple Phases of Liquid Water," *ChemPhysChem* (2008). https://doi.org/10.1002/cphc.200800639.

16. J. D. Bernal and R. H. Fowler, "A Theory of Water and Ionic Solution, with Particular Reference to Hydrogen and Hydroxyl Ions," *The Journal of Chemical Physics* (1933). https://doi.org/10.1063/1.1749327.

17. Henry S. Frank and Wen Yang Wen, "Ion–Solvent Interaction. Structural Aspects of Ion–Solvent Interaction in Aqueous Solutions: A Suggested Picture of Water Structure," *Discussions of the Faraday Society* (1957). https://doi.org/10.1039/DF9572400133.

18. John De Poorter, "An Improved Interstitial-Ice Model for Pure Liquid Water," *SciPost Physics Proceedings* (2020).

19. J. B. Paul, C. P. Collier, R. J. Saykally, J. J. Scherer and A. O'Keefe, "Direct Measurement of Water Cluster Concentrations by Infrared Cavity Ringdown Laser Absorption Spectroscopy," *Journal of Physical Chemistry A* (1997). https://doi.org/10.1021/jp971216z.

20. Kouki Oka, Toshimichi Shibue, Natsuhiko Sugimura, Yuki Watabe, Bjorn Winther-Jensen and Hiroyuki Nishide, "Long-Lived Water Clusters in Hydrophobic Solvents Investigated by Standard NMR Techniques," *Scientific Reports* (2019). https://doi.org/10.1038/s41598-018-36787-1.

21. M. F. Chaplin, "A Proposal for the Structuring of Water," *Biophysical Chemistry* (2000). https://doi.org/10.1016/S0301-4622(99)00142-8.

22. J. A. Sellberg, C. Huang, T. A. McQueen, N. D. Loh, H. Laksmono, D. Schlesinger et al., "Ultrafast X-Ray Probing of Water Structure below the Homogeneous Ice Nucleation Temperature," *Nature* (2014). https://doi.org/10.1038/nature13266.

23. F. Perakis, K. Amann-Winkel, F. Lehmköhler et al., "Diffusive Dynamics during the High-to-Low Density Transition in Amorphous Ice," *Proceedings of the National Academy of Sciences of the United States of America* (2017). https://doi.org/10.1073/pnas.1705303114.

24. H. E. Stanley and F. Mallamace, "Experimental Tests of the Liquid–Liquid Phase Transition Hypothesis," in *Water: Fundamentals as the Basis for Understanding the Environment and Promoting Technology* (2015). https://doi.org/10.3254/978-1-61499-507-4-1.

25. P. H. Handle, T. Loerting and F. Sciortino, "Supercooled and Glassywater: Metastable Liquid(s), Amorphous Solid(s), and a No-Man's Land," *Proceedings of the National Academy of Sciences of the United States of America* (2017). https://doi.org/10.1073/pnas.1700103114.

Chapter 5

1. Yujin Tong, Tobias Kampfrath and R. Kramer Campen, "Experimentally Probing the Libration of Interfacial Water: The Rotational Potential of Water Is Stiffer at the Air/Water Interface than in Bulk Liquid," *Physical Chemistry Chemical Physics* (2016). https://doi.org/10.1039/c6cp01004k.

2. For a review of the methods and results of sum-frequency generation spectroscopy, see Fujie Tang, Tatsuhiko Ohto, Shumei Sun, Jérémy R. Rouxel, Sho Imoto, Ellen H. G. Backus et al., "Molecular Structure and Modeling of Water–Air and Ice–Air Interfaces Monitored by Sum-Frequency Generation," *Chemical Reviews* (2020). https://doi.org/10.1021/acs.chemrev.9b00512.

3. Flaviu S. Cipcigan, Vlad P. Sokhan, Andrew P. Jones, Jason Crain and Glenn J. Martyna, "Hydrogen Bonding and Molecular Orientation at the Liquid–Vapour Interface of Water," *Physical Chemistry Chemical Physics* (2015). https://doi.org/10.1039/c4cp05506c.

4. Simone Pezzotti, Alessandra Serva and Marie Pierre Gaigeot, "2D-HB-Network at the Air–Water Interface: A Structural and Dynamical Characterization by Means of Ab Initio and Classical Molecular Dynamics Simulations," *Journal of Chemical Physics* (2018). https://doi.org/10.1063/1.5018096.

5. J. Joly, "The Phenomena of Skating and Professor J. Thomson's Thermodynamic Relation," *The Scientific Proceedings of the Royal Dublin Society* (1886).

6. "The Mechanism of Sliding on Ice and Snow," *Proceedings of the Royal Society of London. Series A. Mathematical and Physical Sciences* (1939). https://doi.org/10.1098/rspa.1939.0104.

7. Astrid Döppenschmidt, Michael Kappl and Hans Jürgen Butt, "Surface Properties of Ice Studied by Atomic Force Microscopy," *Journal of Physical Chemistry B* (1998). https://doi.org/10.1021/jp981396s.

8. M. Alejandra Sánchez, T. Kling, T. Ishiyama, M. J. Van Zadel, P. J. Bisson, M. Mezger et al., "Experimental and Theoretical Evidence for Bilayer-by-Bilayer Surface Melting of Crystalline Ice," *Proceedings of the National Academy of Sciences of the United States of America* (2017). https://doi.org/10.1073/pnas.1612893114.

9. Bart Weber, Yuki Nagata, Stefania Ketzetzi, Fujie Tang, Wilbert J. Smit, Huib J. Bakker et al., "Molecular Insight into the Slipperiness of Ice," *Journal of Physical Chemistry Letters* (2018). https://doi.org/10.1021/acs.jpclett.8b01188.

10. Xi Zhang, Yongli Huang, Zengsheng Ma, Yichun Zhou, Weitao Zheng, Ji Zhou et al., "A Common Supersolid Skin Covering Both Water and Ice," *Physical Chemistry Chemical Physics* 16:42 (2014): 22987–22994. https://doi.org/10.1039/C4CP02516D.

11. Qianbin Wang, Xi Yao, Huan Liu, David Quéré and Lei Jiang, "Self-Removal of Condensed Water on the Legs of Water Striders," *Proceedings of the National Academy of Sciences of the United States of America* (2015). https://doi.org/10.1073/pnas.1506874112.

12. Yu I. Tarasevich, "State and Structure of Water in Vicinity of Hydrophobic Surfaces," *Colloid Journal* (2011). https://doi.org/10.1134/S1061933X11020141.

13. Jian-ming Zheng and Gerald H. Pollack, "Long-Range Forces Extending from Polymer-Gel Surfaces," *Physical Review E—Statistical Physics, Plasmas, Fluids, and Related Interdisciplinary Topics* (2003). https://doi.org/10.1103/PhysRevE.68.031408.

14. K. Green and T. Otori, "Direct Measurements of Membrane Unstirred Layers," *The Journal of Physiology* (1970). https://doi.org/10.1113/jphysiol.1970.sp009050.

15. H. H. Mollenhauer and D. J. Morre, "Structural Compartmentation of the Cytosol: Zones of Exclusion, Zones of Adhesion, Cytoskeletal and Intercisternal Elements," *Sub-Cellular Biochemistry* (1978). https://doi.org/10.1007/978-1-4615-7942-7_7.

16. Gerald Pollack, *Fourth Phase of Water: Beyond Solid, Liquid and Vapor* (Ebner and Sons, 2013).

17. Daniel C. Elton, Peter D. Spencer, James D. Riches and Elizabeth D. Williams, "Exclusion Zone Phenomena in Water—A Critical Review of Experimental Findings and Theories," arxiv.org/pdf/1909.06822.pdf.

18. *EdgeScience Magazine* 16 (November 2013): 18.

19. For the full story of polywater, see Philip Ball, *H2O: A Biography of Water* (Orion, 1998), chapter 10.

20. Yixing Chen, Halil I. Okur, Nikolaos Gomopoulos, Carlos Macias-Romero, Paul S. Cremer, Poul B. Petersen et al., "Electrolytes Induce Long-Range Orientational Order and Free Energy Changes in the H-Bond Network of Bulk Water," *Science Advances* (2016). https://doi.org/10.1126/sciadv.1501891.

21. Royce K. Lam, Alice H. England, Jacob W. Smith, Anthony M. Rizzuto, Orion Shih, David Prendergast et al., "The Hydration Structure of Dissolved Carbon Dioxide from X-Ray Absorption Spectroscopy," *Chemical Physics Letters* (2015). https://doi.org/10.1016/j.cplett.2015.05.039.

Chapter 6

1. Damien Laage, Thomas Elsaesser and James T. Hynes, "Water Dynamics in the Hydration Shells of Biomolecules," *Chemical Reviews* (2017). https://doi.org/10.1021/acs.chemrev.6b00765.

72. Pavel Jungwirth, "Biological Water or Rather Water in Biology?," *Journal of Physical Chemistry Letters* (2015). https://doi.org/10.1021/acs.jpclett.5b01143.

3. Jessica M. J. Swanson, C. Mark Maupin, Hanning Chen, Matt K. Petersen, Jiancong Xu, Yujie Wu et al., "Proton Solvation and Transport in Aqueous and Biomolecular Systems: Insights from Computer Simulations," *Journal of Physical Chemistry B* (2007). https://doi.org/10.1021/jp070104x.

4. Bobo Feng, Robert P. Sosa, Anna K. F. Mårtensson, Kai Jiang, Alex Tong, Kevin D. Dorfman et al., "Hydrophobic Catalysis and a Potential Biological Role of DNA Unstacking Induced by Environment Effects," *Proceedings of the National Academy of Sciences of the United States of America* (2019). https://doi.org/10.1073/pnas.1909122116.

5. Benoît Roux, "Ion Channels and Ion Selectivity," *Essays in Biochemistry* (2017). https://doi.org/10.1042/EBC20160074.

6. Kevin L. Schey, Zhen Wang, Jamie L. Wenke and Ying Qi, "Aquaporins in the Eye: Expression, Function, and Roles in Ocular Disease," *Biochimica et Biophysica Acta—General Subjects* (2014). https://doi.org/10.1016/j.bbagen.2013.10.037.

7. Justin L. L. Wilson, Carlos A. Miranda and Mark A. Knepper, "Vasopressin and the Regulation of Aquaporin-2," *Clinical and Experimental Nephrology* (2013). https://doi.org/10.1007/s10157-013-0789-5.

8. Eszter Ducza, Adrienn Csányi and Róbert Gáspár, "Aquaporins during Pregnancy: Their Function and Significance," *International Journal of Molecular Sciences* (2017). https://doi.org/10.3390/ijms18122593.

9. Di Zhang, Kaiqi Zhang, Weiheng Su, Yuan Zhao, Xin Ma, Gong Qian et al., "Aquaporin-3 Is Down-Regulated in Jejunum Villi Epithelial Cells during Enterotoxigenic *Escherichia coli*–Induced Diarrhea in Mice," *Microbial Pathogenesis* (2017). https://doi.org/10.1016/j.micpath.2017.04.031.

10. Christophe Maurel, Yann Boursiac, Doan Trung Luu, Véronique Santoni, Zaigham Shahzad and Lionel Verdoucq, "Aquaporins in Plants," *Physiological Reviews* (2015). https://doi.org/10.1152/physrev.00008.2015.

11. Ranganathan Kapilan, Maryam Vaziri and Janusz J. Zwiazek, "Regulation of Aquaporins in Plants under Stress," *Biological Research* (2018). https://doi.org/10.1186/s40659-018-0152-0.

12. Alexander G. Volkov, Tejumade Adesina, Vladislav S. Markin and Emil Jovanov, "Kinetics and Mechanism of *Dionaea muscipula* Trap Closing," *Plant Physiology* (2008). https://doi.org/10.1104/pp.107.108241.

13. Steven Footitt, Rachel Clewes, Mistianne Feeney, William E. Finch-Savage and Lorenzo Frigerio, "Aquaporins Influence Seed Dormancy and Germination in Response to Stress," *Plant Cell and Environment* (2019). https://doi.org/10.1111/pce.13561.

14. "Cavitation and Embolism in Plants: Literature Review," *Australian Journal of Basic and Applied Sciences* (2018). https://doi.org/10.22587/ajbas.2018.12.5.1.

15. Sukrit Ranjan, Zoe R. Todd, Paul B. Rimmer, Dimitar D. Sasselov and Andrew R. Babbin, "Nitrogen Oxide Concentrations in Natural Waters on Early Earth," *Geochemistry, Geophysics, Geosystems* (2019). https://doi.org/10.1029/2018GC008082.

16. Inho Nam, Jae Kyoo Lee, Hong Gil Nam and Richard N. Zare, "Abiotic Production of Sugar Phosphates and Uridine Ribonucleoside in Aqueous Microdroplets," *Proceedings of the National Academy of Sciences of the United States of America* (2017). https://doi.org/10.1073/pnas.1714896114.

17. Albert Szent-Györgyi, "Biology and Pathology of Water," *Perspectives in Biology and Medicine* 14:2 (1971): 239–249. doi:10.1353/pbm.1971.0014.

Index